近 世 代 数

王栓宏　编著

科 学 出 版 社

北 京

内 容 简 介

本书主要介绍了群胚(groupoid)、群(group)、环(ring)和模(module)的基本概念和理论，并特别介绍了与这些概念相关的国际前沿研究课题和应用. 本书内容由浅入深，结合双语课程的特点，在编写方法上对如何组织双语教材进行了有益的探索.

本书可供高等学校数学及相关专业高年级本科生和高校教师从事双语课程教学时阅读和参考.

图书在版编目(CIP)数据

近世代数/王栓宏编著. —北京：科学出版社，2013
ISBN 978-7-03-036867-6

Ⅰ. ①近… Ⅱ. ①王… Ⅲ. ①抽象代数—双语教学—教材 Ⅳ. ①O153

中国版本图书馆 CIP 数据核字(2013) 第 040560 号

责任编辑：顾 艳 尚 雁／责任校对：张怡君
责任印制：徐晓晨／封面设计：许 瑞

科学出版社出版
北京东黄城根北街 16 号
邮政编码：100717
http://www.sciencep.com
北京凌奇印刷有限责任公司印刷
科学出版社发行 各地新华书店经销
*
2013 年 3 月第 一 版 开本：B5(720×1000)
2025 年 1 月第六次印刷 印张：5 3/4
字数：120 000

定价：29.00 元
(如有印装质量问题，我社负责调换)

前　言

目前, 国内许多大学已经意识到本科教学国际化的重要性, 教学计划中也设置了双语课程, 于是, 如何开展与评估双语教学, 尤其是如何编写双语教材, 学生手上应该有什么样的教材就成为一个很大的问题, 有待探索与研究.

本科教学的国际化最终的落脚点是本科生的国际化, 包括出国交换学习、参加国际会议、在国内听外国专家的各种讲座与学术报告. 不论是研究生还是老师, 与外国同行交流专业知识时, 概念和性质的叙述非常重要, 必须用英语完成; 至于性质的证明则是一种深入的研究过程. 根据作者多年来与国外同行专家合作研究以及参加各种国际会议的经验和体会, 本科生双语教材编写的主要目的是使学生学会应用英语来描述和掌握概念、性质、例子和应用, 而淡化应用英语来证明性质、定理等的研究过程, 也就是说, 我们可用中文给出所有证明过程, 让学生真正领会抽象的概念并会用严密的数学推导来证明相应的性质、定理等, 避免由于对英语语言的误解而没有理解代数专业知识的探究.

正是基于这样的考虑, 本教材在内容与写作上有如下三个特点: 第一, 内容处理上兼顾了与所讲概念相关的国际前沿研究性问题; 第二, 增加了理论的应用内容; 第三, 用英语描述概念、性质、定理、例子、应用与习题等, 所有性质、定理等的证明用中文叙述. 我们相信, 通过这样的双语教材的学习, 本科生与国际同行专家交流本方向的专业知识时就会得心应手.

在完成这本书稿的过程中, 博士生王伟、郭双建、王圣祥和张晓辉做了大量校稿和打印工作; 本书的出版得到了东南大学双语教材项目的资助, 在此一并致谢.

书中难免有不足之处, 恳请读者批评指正.

作　者

2012 年 12 月于南京

目　　录

第1章　群胚 (Groupoids)

本章主要介绍群胚 (groupoid) 的概念. 这一概念是由 Heinrich Brandt 在 1926 年最先引进的, 它的深层次研究会涉及弱 Hopf 代数 (weak Hopf algebra) 与弱乘子 Hopf 代数 (weak multiplier Hopf algebra) 理论的建立. 为了构造群胚, 我们首先介绍等价关系与集合分类.

1.1　等价关系 (Equivalencc Relations)

本节介绍关系 (relation) 且给出例子, 进一步研究等价关系 (equivalence relation).

Definition 1.1.1　A **relation**(关系) on a nonempty set A is a collection of ordered pairs of elements of A. In other words, it is a subset of the Cartesian product $A^2 = A \times A$.

If S is a set, we will use the symbol "$\sim$" to denote either an abstract relation or a specific relation for which there is no standard notation. For $a, b \in S$, we will write $a \sim b$, not $(a, b) \in \sim$, to indicate that a and b are related.

Definition 1.1.2　Let $\sim$ be a relation of a set S.

(1) We say that $\sim$ is **reflexive**(自反性) provided for all $a \in S$, it implies $a \sim a$.

(2) We say that $\sim$ is **symmetric**(对称性) provided for all $a, b \in S$, $a \sim b$ implies $b \sim a$.

(3) We say that $\sim$ is **transitive** (传递性) provided for all $a, b, c \in S$, if $a \sim b$ and $b \sim c$, then $a \sim c$.

Example 1.1.3　(1) Consider the relation "$<$" on $\mathbb{R}$. It is easy to check that "$<$" is transitive but not reflexive and symmetric.

(2) Consider the relation "$\leqslant$" on $\mathbb{R}$. It is easy to check that "$\leqslant$" is transitive and reflexive but not symmetric.

(3) Define the relation $\sim$ as follows: $a, b \in \mathbb{Q}$, if $|a - b| < 1$, then $a \sim b$. It is easy to show that $\sim$ is symmetric and reflexive but not transitive.

Definition 1.1.4　A relation $\sim$ on a set S is called an **equivalence relation** provided $\sim$ is reflexive, symmetric, and transitive.

Example 1.1.5　(1) For $x, y \in \mathbb{R}$, define $x \sim y$ to mean that $x - y \in \mathbb{Z}$. It can be showed that $\sim$ is an equivalence relation on $\mathbb{R}$.

(2) Define the "square" relation on $\mathbb{R}$ to mean that $x^2 = y^2$. Then the square relation is an equivalence relation.

(3) For $a, b \in \mathbb{Z}$, define $a \sim b$ to mean that a divides b. Since 0 does not divide 0, $\sim$ is not an equivalence relation.

Definition 1.1.6　Let n be a positive integer. For integers a and b, we say that a is **congruent to** b **modulo** n(a 与 b 同余模 n), and write $a \equiv b \pmod{n}$, provided $a - b$ is divisible by n.

The following statements are various ways to say $a \equiv b \pmod{n}$, that is, the statements are equivalent:

(i) $a \equiv b \pmod{n}$;

(ii) $a - b = kn$ for some integer k;

(iii) $a = kn + b$ for some integer k.

Theorem 1.1.7　Congruence modulo n is an equivalence relation on $\mathbb{Z}$.

证明　首先来验证反身性. 设 a 为任一整数. 由 $a - a = 0$ 和 0 可被 n 整除, 知 $a \equiv a \pmod{n}$ 成立.

下证对称性. 设 a 和 b 为整数, 则知必存在整数 k 使得 $a - b = kn$, 此时有 $b - a = -(a - b) = -(kn) = (-k)n$. 故 n 整除 $b - a$, 即 $b \equiv a \pmod{n}$.

再证传递性. 设 $a \equiv b \pmod{n}$, $b \equiv c \pmod{n}$, 则必存在整数 k, k' 使得 $a - b = kn$, $b - c = k'n$. 于是 $a - c = (k + k')n$. 故 n 整除 $a - c$, 即 $a \equiv c \pmod{n}$. □

1.2　等价类 (Equivalence Classes)

本节利用等价关系对非空集合进行分类, 同时讨论这两个概念的等价性.

Definition 1.2.1　Let $\sim$ be an equivalence relation on a set S. For each $a \in S$, we define the **equivalence class** (等价类) of a, denoted by $\overline{a}$, to be the set

$$\overline{a} = \{b \in S \mid b \sim a\}.$$

Example 1.2.2 For $x, y \in \mathbb{R}$, define $x \sim y$ to mean that $|x| = |y|$. Then for every $a \in \mathbb{R}$, we have

$$\overline{a} = \{a, -a\}.$$

Theorem 1.2.3 Let $\sim$ be an equivalence relation on the set S. For $a, b \in S$, the following statements are equivalent:

(i) $\overline{a} = \overline{b}$;

(ii) $a \sim b$;

(iii) $a \in \overline{b}$;

(iv) $\overline{a} \cap \overline{b} \neq \varnothing$.

证明 (i)⇒(ii): 易知 $b \in \overline{b} = \overline{a}$, 即 $a \sim b$.

(ii)⇒(iii): 显然.

(iii)⇒(iv): 由 $a \in \overline{b}$ 和 $a \in \overline{a}$, 即可知 $\overline{a} \cap \overline{b} \neq \varnothing$.

(iv)⇒(i): 设 $c \in \overline{a} \cap \overline{b}$, 则 $c \sim a$ 且 $c \sim b$. 再任取 $d \in \overline{a}$, 知 $d \sim a$. 于是由 $\sim$ 为等价关系, 满足传递性, 且 $c \sim a$ 和 $d \sim a$ 可知, $d \sim c$. 又由 $c \sim b$, 故 $d \in \overline{b}$, 即 $\overline{a} \subseteq \overline{b}$.

同理可知 $\overline{a} \supseteq \overline{b}$. 于是 $\overline{a} = \overline{b}$. □

Equivalence classes for congruence mod n are also called **congruence classes**(同余类). Let a be an integer. By the definition of an equivalence class we have

$$\overline{a} = \{x \in \mathbb{Z} \mid x \equiv a \bmod(n)\} = \{x \in \mathbb{Z} \mid x - a = kn \text{ for some integer } k\}.$$

For the given n, we denote by $\mathbb{Z}_n$ the set of all congruence classes of $\mathbb{Z}$ for the relation congruence mod n. Thus, $\mathbb{Z}_n = \{\overline{a} \mid a \in \mathbb{Z}\}$.

Theorem 1.2.4 For every positive integer n, $\mathbb{Z}_n = \{\overline{0}, \overline{1}, \cdots, \overline{n-1}\}$.

证明 设 $a \in \mathbb{Z}$, 则由带余除法公理知必存在整数 q, r, 使得 $a = qn + r$ 且 $0 \leqslant r < n$. 故 $a - r = qn$, 即 $a \in \overline{r}$. 由定理 1.2.3 可知,$\overline{a} = \overline{r}$. 从而每个小于 0 或大于 $n-1$ 的数都可以被 $0, 1, \cdots, n-1$ 之中的某数表达出来. 故命题成立. □

Corollary 1.2.5 Let n be a positive integer. For every $a, b \in \mathbb{Z}$, the following statements are equivalent:

(i) In $\mathbb{Z}_n$, $\overline{a} = \overline{b}$;

(ii) $a \equiv b \pmod{n}$, that is, n divides $a - b$;

(iii) $a \in \overline{b}$;

(iv) $\overline{a} \cap \overline{b} \neq \varnothing$.

证明 结合定理 1.2.3 和定理 1.2.4 即可得. □

We know that the set of all equivalence classes of S under $\sim$ is defined to be the set of all subsets of S which are equivalence classes of S under $\sim$, and is denoted by $S/\sim$. The map $x \mapsto \overline{x}$ is sometimes referred to as the **canonical projection**(标准投射).

Definition 1.2.6　Let S be a nonempty set. If S can be represented as a union of some subsets and these subsets are disjoint sets with each other, i.e. $S = \bigcup_{i=1}^{n} S_i$, where $S_i \subseteq S$, $n = 1, 2, \cdots, \infty$, $S_i \cap S_j = \varnothing$ for $i \neq j$, then every S_i is called a **class** of S, and $\{S_i \mid i = 1, 2, \cdots, n\}$ is called a **partition** (划分) of S.

Let S be a set with an equivalence relation $\sim$. We will see that for every equivalence relation $\sim$, the set of all equivalence classes of S under $\sim$ is a partition of S, and this correspondence is a bijection between the set of equivalence relations on S and the set of partitions of S (consisting of nonempty sets).

Theorem 1.2.7　A partition of a nonempty set S decided an equivalence relation $\sim$ in S.

证明　设 $S=\bigcup_{i=1}^{n} S_i$, 其中 $S_i \subseteq S$, $n=1,2,\cdots,\infty$, 且若 $i \neq j$, S_i 与 S_j 非交. 定义 S 中的关系 $\sim$ 如下:

$$a \sim b \Longleftrightarrow a\text{与}b\text{同在一类, 即}a, b\text{同属于某个}S_i.$$

下证 $\sim$ 为一个等价关系.

对称性: 若 $a \sim b$, 则存在 i, 使得 $a, b \in S_i$. 显然有 $b \sim a$.

反身性: 若 $a \in S$, 则必存在 i, 使得 $a \in S_i$. 即 $a \sim a$.

传递性: 若 $a \sim b$, $b \sim c$, 则存在 i, j, 使得 $a, b \in S_i$, $b, c \in S_j$, 即 $b \in S_i \cap S_j$. 从而必有 $i = j$. 即 $a, b, c \in S_i$. 于是 $a \sim c$.

综上可知, $\sim$ 为等价关系. □

Theorem 1.2.8　An equivalence relation $\sim$ in a nonempty set S decided a partition of S.

证明　设 $\sim$ 为 S 中的一个等价关系, 则易知 $S = \bigcup_{a \in S} \overline{a}$, 其中 $\overline{a}$ 为 a 的等价类. 由定理 1.2.3, 可知若 a 与 b 不等价, 则 $\overline{a}$ 与 $\overline{b}$ 无交. 从而可得 $S = \bigcup_{a_i \in S} \overline{a_i}$, 其中 a_i 取遍 S 中关于 $\sim$ 的等价类的代表元. □

Then we can get the following

Corollary 1.2.9　For a given set S, $\sim$ is an equivalence relation if and only if it produces a partition.

Definition 1.2.10 Let $\sim$ be an equivalence relation on nonempty set S. The set of all equivalence classes in S is called the **quotient set** (商集) of S relative to $\sim$, and is denoted by $S/\sim$ or $\overline{S}$, that is,

$$S/\sim = \{\overline{a} \mid a \in A\} \subseteq P(S),$$

where $P(S)$ is the power set of S.

Define a canonical projection by

$$v : S \to S/\sim \text{ via } a \mapsto \overline{a},$$

which is a surjection.

1.3 群胚 (Groupoids)

本节主要介绍群胚 (groupoid) 的概念, 并给出构造群胚的方法, 同时指出与之相关的国际前沿研究的问题. 见文献 [4]、[5]、[9]、[10] 和 [12].

Definition 1.3.1 A **groupoid** (群胚) is a set G with a unary operation $^{-1}: G \to G$ and a partial function $*: G \times G \to G$ (here $*$ is **not** necessarily defined for all possible pairs of G-elements) such that the following axiomatic properties hold: for every $a, b, c \in G$

(i) **Associativity**(结合性): If $a * b$ and $b * c$ are defined, then $(a * b) * c$ and $a * (b * c)$ are defined and equal. Conversely, if either of these last two expressions is defined, then so is the other (and again they are equal);

(ii) **Inverse**(可逆性): $a^{-1} * a$ and $a * a^{-1}$ are always defined;

(iii) **Identity**(单位性): If $a * b$ is defined, then $a * b * b^{-1} = a$, and $a^{-1} * a * b = b$.

From these axioms, it is easy to get $(a^{-1})^{-1} = a$ and that if $a * b$ is defined, then $(a * b)^{-1} = b^{-1} * a^{-1}$.

Remark 1.3.2 Let G be a groupoid. When G is a finite set, we say that G is a **finite groupoid** (有限群胚). When G is an infinite set, we call G an **infinite groupoid** (无限群胚).

Let G be a groupoid. It is a set with a distinguished subset of pairs (p, q) in $G \times G$ for which the product pq in G is defined. This product is associative, in the appropriate sense. The product pq is only defined when the so-called source $s(p)$ of the element p is equal to the target $t(q)$ of the element q. The source and target maps are defined from G to the set of units and this set can (and will) be considered as a subset of G.

Examples 1.3.3 The simplest example of a groupoid is obtained from an equivalence relation $\sim$ on a set X. The elements of the groupoid G are pairs (y, x) with $x, y \in X$ and $x \sim y$. The set of units is X and the source and target maps are seperately given by

$$s(y, x) = x \quad \text{and} \quad t(y, x) = y,$$

for (y, x) in G. The set of units is considered as a subset of G via the map $x \to (x, x)$. The product of (z, y) with (y, x) is (z, x) when $x, y, x \in X$, $x \sim y$ and $y \sim z$.

国际前沿研究动态

(1) 当 G 是一个有限群胚时, 由此可以定义一个弱 Hopf 代数 (weak Hopf algebra), 研究见文献 [2]、[3]、[6]~[8].

(2) 当 G 是一个无限群胚时, 由此可以定义一个弱乘子 Hopf 代数 (weak multiplier Hopf algebra), 研究见文献 [1]、[11]、[13]~[18].

参考文献

[1] Abe E. Hopf Algebras. New York: Cambridge University Press, 1977.

[2] Böhm G, Nill F, Szlachányi K. Weak Hopf algebras I: integral theory and C^*-structure. Journal of Algebra, 1999, 221(2): 385-438.

[3] Böhm G, Szlachányi K. Weak Hopf algebras II: representation theory, dimensions and the Markov trace. Journal of Algebra, 2000, 233(1): 156-212.

[4] Brown R. From groups to groupoids: a brief survey. Bull. London Math. Soc., 1987, 19: 113-134.

[5] Higgins P J. Notes on Categories and Groupoids// Halmos P R, et al., ed. Van Nostrand Reinhold Mathematical Studies. Vol.32. London: Van Nostrand Reinhold, 1971.

[6] Nikshych D. On the structure of weak Hopf algebras. Adv. Math., 2002, 170: 257-286.

[7] Nikshych D, Vainerman L. Algebraic versions of a finite dimensional quantum groupoid. Lecture Notes in Pure and Applied Mathematics, 2000, 209: 189-221.

[8] Nikshych D, Vainerman L. Finite quantum groupiods and their applications. New Directions in Hopf Algebras, 2002, 43: 211-262.

[9] Paterson A. Groupoids, Inverse Semigroups and Their Operator Algebras. Boston: Birkhauser, 1999.

[10] Renault J. A Groupoid Approach to C^*-algebras//Morel, et al. Lecture Notes in Mathematics. Vol.793. Berlin: Springer Verlag, 1980.

[11] Sweedler M. Hopf Algebras. New York: Benjamin, 1969.

[12] Vainerman L. Locally compact quantum groups and groupoids. IRMA Lectures in Mathematics and Theoretical Physics 2. Proceedings of a meeting in Strasbourg, de Gruyter, 2002.

[13] Van Daele A. Multiplier Hopf algebras. Trans. Am. Math. Soc., 1994, 342(2): 917-932.

[14] Van Daele A. An algebraic framework for group duality. Adv. in Math., 1998, 140: 323-366.

[15] Van Daele A, Wang S H. Weak multiplier Hopf algebras. Preliminaries, motivation and basic examples. Preprint University of Leuven and Southeast University of Nanjing (2012). Arxiv:1210.3954v1 [math.RA]. To appear in the proceedings of the conference .Operator Algebras and Quantum Groups (Warsaw, September 2011), series .Banach Center Publications..

[16] Van Daele A, Wang S H. Weak multiplier Hopf algebras I: The main theory. Preprint University of Leuven and Southeast University of Nanjing (2012). Arxiv: 1210.4395v1[math.RA].

[17] Van Daele A, Wang S H. Weak multiplier Hopf algebras II: The source and target algebras. University of Leuven and Southeast University of Nanjing (in preparation).

[18] Van Daele A, Wang S H. Weak multiplier Hopf algebras III: Integrals and duality. University of Leuven and Southeast University of Nanjing (in preparation).

习 题

1. For $A, B \in P(Z)$, define $A \sim B$ to mean that $A \cap B = \varnothing$ (Recall that $P(Z)$ is the power set of Z),

(i) Prove or disprove that $\sim$ is reflexive;

(ii) Prove or disprove that $\sim$ is symmetric;

(iii) Prove or disprove that $\sim$ is transitive.

2. For $(a,b),(c,d) \in \mathbb{R}^2$, define $(a,b) \sim (c,d)$ to mean that $2a-b=2c-d$. Prove that $\sim$ is an equivalence relation on $\mathbb{R}^2$.

3. Define a function $f:\mathbb{R}\to\mathbb{R}$ by $f(x)=x^2+1$. For $a,b\in\mathbb{R}$, define $a\sim b$ to mean that $f(a)=f(b)$.

(i) Prove that $\sim$ is an equivalence relation on $\mathbb{R}$;

(ii) List all elements in the set $\{x\in\mathbb{R}|x\sim 3\}$.

4. Describe the set of all integers x such that $x\equiv 4 \pmod 9$ and use the description to list all integers x such that $-36\leqslant x\leqslant 36$ and $x\equiv 4 \pmod 9$.

5. Let m and n be positive integers such that m divides n. Prove that for all integers a and b, if $a\equiv b \pmod n$, then $a\equiv b \pmod m$.

6. Let $\mathbb{R}^*$ denote the set of all nonzero real numbers and let $\mathbb{Q}^*$ denote the set of all nonzero rational numbers. For $a,b\in\mathbb{R}^*$, define $a\sim b$ to mean that $a/b\in\mathbb{Q}^*$. Prove that $\sim$ is an equivalence relation, and prove each of the following:

(i) $\overline{\sqrt{3}}=\overline{\sqrt{12}}$;

(ii) $\overline{\sqrt{3}}\cap\overline{\sqrt{6}}=\varnothing$;

(iii) $\overline{\sqrt{8}}\neq\overline{\sqrt{12}}$;

(iv) $x=3$ is the solution to the equation $\overline{x\sqrt{2}}=\overline{2\sqrt{2}}$.

7. In $\mathbb{Z}_9$, prove or disprove

(i) $\overline{32}=\overline{50}$;

(ii) $\overline{-33}=\overline{75}$;

(iii) $\overline{-16}=\overline{-37}$.

8. For a positive integer n, set

$$\mathbb{Z}_{(n)}=\{\overline{a}\in\mathbb{Z}_n \mid \gcd(a,n)=1\}.$$

Thus, for example, $\mathbb{Z}_{(10)}=\{\overline{1},\overline{3},\overline{7},\overline{9}\}$.

(i) Prove that the set $\mathbb{Z}_{(n)}$ is well-defined, that is, prove that for all integers a_1 and a_2, if $\overline{a_1}=\overline{a_2}$ in $\mathbb{Z}_{(n)}$ and $\overline{a_1}\in\mathbb{Z}_{(n)}$, then $\overline{a_2}\in\mathbb{Z}_{(n)}$;

(ii) Prove that for all integers a and b, if $\overline{a},\overline{b}\in\mathbb{Z}_{(n)}$, then $\overline{ab}\in\mathbb{Z}_{(n)}$.

第 2 章　群 (Groups)

本章主要介绍群 (group) 的概念 (最先由 E.Galois 引进) 及其结构性质, 涉及子群 (subgroup)、正规子群 (normal subgroup)、对称群 (symmetric group)、群同态 (group homomorphism) 和商群 (quotient group). 它的深层次研究会涉及量子群 (quantum group) 与乘子 Hopf 代数 (multiplier Hopf algebra) 理论的建立.

2.1　群　概　念

本节主要介绍群的定义并给出群的简单例子.

Definition 2.1.1　A **binary operation**(二元运算) on a set G is a function $*: G \times G \to G$.

In more details, an operation assigns an element $*(x, y)$ in G to each ordered pair (x, y) of elements in G. It is more natural to write $x * y$ instead of $*(x, y)$; thus, composition of functions is the function $(f, g) \mapsto g \circ f$, while multiplication, addition, and subtraction are respectively the functions $(x, y) \mapsto xy$, $(x, y) \mapsto x + y$, and $(x, y) \mapsto x - y$. The examples of composition and subtraction show why we want ordered pairs, for $x * y$ and $y * x$ may be distinct. As function, each operation is single-valued; when one says this explicitly, it is usually called the **law of substitution**:

$$\text{if } x = x' \text{ and } y = y', \text{ then } x * y = x' * y'.$$

Definition 2.1.2　A **group**(群) $(G, *, e)$ is a nonempty set G equipped with a binary operation $*$ and a special element $e \in G$ called the **identity**(单位元), such that

(1) **Associativity**(结合性) holds: for every $a, b, c \in G$,

$$a * (b * c) = (a * b) * c;$$

(2) **Left Identity**(左单位性) holds: $e * a = a$ for all $a \in G$, i.e., e is a left identity;

(3) **Left inverse** (左逆性) holds: for every $a \in G$, there is $a' \in G$ with $a' * a = e$, i.e., a' is a left inverse of a.

Remark 2.1.3 We can refer to G in place of a group $(G, *, e)$, provided that no confusion will result.

Example 2.1.4 (1) The set of integers under addition forms an infinite additive group, $(\mathbb{Z}, +, 0)$.

(2) The set of nonzero real numbers under multiplication forms an infinite multiplicative group, $(\mathbb{R} \setminus \{0\}, \times, 1)$.

(3) The complex numbers under the operation of addition forms a group, $(\mathbb{C}, +, 0)$.

(4) The set

$$S = \{x + y\sqrt{2} | x, y \in \mathbb{Z}, \text{ where } (x, y) \neq (0, 0)\}$$

under multiplication forms a group, $(S, \times, 1)$.

Let G be a group. If for all $x, y \in G$, $xy = yx$, we call the group **Abelian or commutative** (阿贝尔或交换的).

Lemma 2.1.5 If $*$ is an associative operation on a set G, then

$$(a * b) * (c * d) = [a * (b * c)] * d,$$

for all $a, b, c, d \in G$.

证明 如果设 $g = a * b$, 那么就有 $(a * b) * (c * d) = g * (c * d) = (g * c) * d = [(a * b) * c] * d = [a * (b * c)] * d$. □

Lemma 2.1.6 If G is a group and $a \in G$ satisfies $a * a = a$, then $a = e$.

证明 由群定义知, 存在 $a' \in G$ 使得 $a' * a = e$. 在 $a * a = a$ 左右两边同时乘以 a', 则等式右边为 e, 而左边为 $a' * (a * a) = (a' * a) * a = e * a = a$, 所以 $a = e$. □

Proposition 2.1.7 Let G be a group with operation $*$ and identity e, then we have

(1) $a * a' = e$ for all $a \in G$;

(2) $a * e = a$ for all $a \in G$;

(3) if $e_0 \in G$ satisfies $e_0 * a = a$ for all $a \in G$, then $e_0 = e$;

(4) if $a, b \in G$ satisfy $b * a = e$, then $b = a'$.

证明 (1) 我们已知 $a' * a = e$, 接下来证明 $a * a' = e$. 由引理 2.1.5,

$$
\begin{aligned}
(a * a') * (a * a') &= [a * (a' * a)] * a' \\
&= (a * e) * a' \\
&= a * (e * a') \\
&= a * a'.
\end{aligned}
$$

再根据引理 2.1.6, $a * a' = e$.

(2) 应用 (1)

$$a * e = a * (a' * a) = (a * a') * a = e * a = a.$$

因此, $a * e = a$.

(3) 我们现在证明一个群只含有唯一的单位元. 如果 $e_0 * a = a$, 对所有的 $a \in G$, 那么特别地 $e_0 * e_0 = e_0$. 再由引理 2.1.6, $e_0 = e$.

(4) 在 (1) 中, 我们证得如果 $a' * a = e$, 则 $a * a' = e$. 那么

$$b = b * e = b * (a * a') = (b * a) * a' = e * a' = a'. \qquad \square$$

From this proposition, we know that a **left inverse** is also a **right inverse**. Thus, we use a^{-1} to denote the inverse of a. The identity element of a group can also be denoted by 1.

We now present a very important group. In high school mathematics, the words of permutation and arrangement are used interchangeably, if the word arrangement is used at all. We draw a distinction between them.

Definition 2.1.8 If X is a set, then a **list** in X is a function $f : \{1, 2, \cdots, n\} \to X$. If a list f in X is a bijection (so that X is now a finite set with $|X| = n$), then f is called an **arrangement** of X.

If f is a list, denote its values $f(i)$ by x_i, where $1 \leqslant i \leqslant n$. Thus, a list in X is merely an n-tuple $(x_1, x_2, \cdots, x_n)$. To say that a list f is injective is to say that there are no repeated coordinates [if $i \neq j$, then $x_i = f(i) \neq f(j) = x_j$]; to say that f is surjective is to say that every $x \in X$ occurs as some coordinate. Thus, an arrangement of X is an n-tuple $(x_1, x_2, \cdots, x_n)$ of all the elements of X with no repetitions. We often omit parentheses and write a list as $x_1, x_2, \cdots, x_n$. For example, there are 27 lists in $X = \{a, b, c\}$ and 6 arrangements:

$$abc; \quad acb; \quad bac; \quad bca; \quad cab; \quad cba.$$

All we can do with such lists is to count the number of them; there are exactly n^n lists and $n!$ arrangements of an n-element set X.

Definition 2.1.9 A **permutation**(置换) of a possibly infinite set X is bijection $\alpha : X \to X$.

Given a finite set X with $|X| = n$, let $\varphi : \{1, 2, \cdots, n\} \to X$ be an arrangement chosen once for all; of course, φ is a bijection. If $f : \{1, 2, \cdots, n\} \to X$ is an arrangement of X, then $f \circ \varphi^{-1} : X \to X$ is a permutation of X. Conversely, if $\alpha : X \to X$ is a permutation of X, then $\alpha \circ \varphi : \{1, 2, \cdots, n\} \to X$ is an arrangement of X. Thus, arrangements and permutations are simply different ways of describing the same thing. The advantage of using permutations, instead of arrangements, is that permutations can be composed and, by Exercises of this chapter, their composite is also a permutation.

If $X = \{1, 2, \cdots, n\}$, then we may use a two-rowed notation to denote a permutation α:

$$\alpha = \begin{pmatrix} 1 & 2 & \cdots & j & \cdots & n \\ \alpha(1) & \alpha(2) & \cdots & \alpha(j) & \cdots & \alpha(n) \end{pmatrix}.$$

Thus, the bottom row is the arrangement $\alpha(1), \alpha(2), \cdots, \alpha(n)$.

Most of the results in this section first appeared in an article of Cauchy in 1815.

Definition 2.1.10 The family of all permutations of a set X, denoted by S_X, is called the **symmetric group** (对称群) on X. When $X = \{1, 2, \cdots, n\}$, S_X is usually denoted by S_n, and it is called the **symmetric group on n letters**(n 元对称群).

Notice that composition in S_3 is not commutative. If

$$\alpha = \begin{pmatrix} 1 & 2 & 3 \\ 2 & 3 & 1 \end{pmatrix} \quad \text{and} \quad \beta = \begin{pmatrix} 1 & 2 & 3 \\ 2 & 1 & 3 \end{pmatrix},$$

then their composites are

$$\alpha \circ \beta = \begin{pmatrix} 1 & 2 & 3 \\ 3 & 2 & 1 \end{pmatrix} \quad \text{and} \quad \beta \circ \alpha = \begin{pmatrix} 1 & 2 & 3 \\ 1 & 3 & 2 \end{pmatrix}.$$

Thus, $\alpha \circ \beta : 1 \mapsto \alpha(\beta(1)) = \alpha(2) = 3$ while $\beta \circ \alpha : 1 \mapsto 2 \mapsto 1$, and so $\alpha \circ \beta \neq \beta \circ \alpha$.

On the other hand, some permutations do commute; for example,

$$\gamma = \begin{pmatrix} 1 & 2 & 3 & 4 \\ 2 & 1 & 3 & 4 \end{pmatrix} \quad \text{and} \quad \delta = \begin{pmatrix} 1 & 2 & 3 & 4 \\ 1 & 2 & 4 & 3 \end{pmatrix}$$

commute, as the reader may check.

Composition in S_X satisfies the cancellation law:

$$\text{if } \gamma \circ \alpha = \gamma \circ \beta, \text{ then } \alpha = \beta.$$

To see this,

$$\begin{aligned} \alpha &= 1_X \circ \alpha \\ &= (\gamma^{-1} \circ r) \circ \alpha \\ &= \gamma^{-1} \circ (\gamma \circ \alpha) \\ &= \gamma^{-1} \circ (\gamma \circ \beta) \\ &= (\gamma^{-1} \circ \gamma) \circ \beta \\ &= 1_X \circ \beta = \beta. \end{aligned}$$

A similar argument shows that

$$\alpha \circ \gamma = \beta \circ \gamma \text{ implies } \alpha = \beta.$$

Aside from being cumbersome, there is a major problem with the two-rowed notation for permutations. It hides the answers to such elementary questions as: Is the square of a permutation the identity? What is the smallest positive integer m so that the m-th power of a permutation is the identity? Can one factor a permutation into simpler permutations? The special permutations introduced below will solve these problems.

Let us first simplify notation by writing $\beta\alpha$ instead of $\beta \circ \alpha$ and (1) instead of 1_X.

Definition 2.1.11 If $\alpha \in S_n$ and $i \in \{1, 2, \cdots, n\}$, then α **fixes** i if $\alpha(i) = i$, and α **moves** i if $\alpha(i) \neq i$.

Definition 2.1.12 Let $i_1, i_2, \cdots, i_r$ be distinct integers in $\{1, 2, \cdots, n\}$. If $\alpha \in S_n$ fixes the other integers (if any) and if

$$\alpha(i_1) = i_2, \quad \alpha(i_2) = i_3, \quad \cdots, \quad \alpha(i_{r-1}) = i_r, \quad \alpha(i_r) = i_1,$$

then α is called an **r-cycle**. One also says that α is a cycle of **length** r.

A 2-cycle interchanges i_1 and i_2 and fixes everything else; 2-cycles are also called **transpositions**. A 1-cycle is the identity, for it fixes every i; thus, all 1-cycles are equal: $(i) = (1)$ for all i.

Consider the permutation

$$\alpha = \begin{pmatrix} 1 & 2 & 3 & 4 & 5 \\ 4 & 3 & 1 & 5 & 2 \end{pmatrix}.$$

The two-rowed notation does not help us recognize that α is, in fact, a 5-cycle: $\alpha(1) = 4, \alpha(4) = 5, \alpha(5) = 2, \alpha(2) = 3$, and $\alpha(3) = 1$. We now introduce new notation: an r-cycle α, as in the definition, shall be denoted by

$$\alpha = (i_1\ i_2\ \cdots\ i_r).$$

For example, the preceding 5-cycle α will be written as $\alpha = (1\ 4\ 5\ 2\ 3)$. The reader may check that

$$\begin{pmatrix} 1 & 2 & 3 & 4 \\ 2 & 3 & 4 & 1 \end{pmatrix} = (1\ 2\ 3\ 4),$$

$$\begin{pmatrix} 1 & 2 & 3 & 4 & 5 \\ 5 & 1 & 4 & 3 & 2 \end{pmatrix} = (1\ 5\ 3\ 4\ 2),$$

and

$$\begin{pmatrix} 1 & 2 & 3 & 4 & 5 \\ 2 & 3 & 1 & 4 & 5 \end{pmatrix} = (1\ 2\ 3).$$

Notice that

$$\beta = \begin{pmatrix} 1 & 2 & 3 & 4 \\ 2 & 1 & 4 & 3 \end{pmatrix}$$

is not a cycle; in fact, $\beta = (1\ 2)(3\ 4)$. The term **cycle** comes from the Greek word for circle. Picture the cycle $(i_1\ i_2\ \cdots\ i_r)$ as a clockwise rotation of the circle. Every i_j can be taken as the "starting point", and so there are r different cycle notations for every r-cycle:

$$(i_1\ i_2\ \cdots\ i_r) = (i_2\ i_3\ \cdots i_r\ i_1) = \cdots = (i_r\ i_1\ i_2\ \cdots\ i_{r-1}).$$

Lemma 2.1.13 Let G be a group.

(1) The **cancellation laws**(消去律) hold: if $a, b, x \in G$, and either $x * a = x * b$ or $a * x = b * x$, then $a = b$.

(2) $(a^{-1})^{-1} = a$ for all $a \in G$.

(3) If $a, b \in G$, then

$$(a * b)^{-1} = b^{-1} * a^{-1}.$$

证明 留给读者作为练习. (提示: 利用引理 2.1.5 和命题 2.1.7)

Definition 2.1.14 If G is a group and if $a \in G$, define the **powers** a^n, for $n \geqslant 1$, inductively:

$$a^1 = a \quad \text{and} \quad a^{n+1} = aa^n.$$

Define $a^0 = e$ and, if n is a positive integer, define

$$a^{-n} = (a^{-1})^n.$$

We let the reader prove that $(a^{-1})^n = (a^n)^{-1}$; this is a special case of the equation in Lemma 2.1.13.

It is easy to check that for a group G, if $a \in G$, and if $m, n \geqslant 1$, then $a^{m+n} = a^m a^n$ and $(a^m)^n = a^{mn}$.

Proposition 2.1.15 Let G be a group, let $a, b \in G$, and let m and n be integers.

(1) If a and b commute, then $(ab)^n = a^n b^n$.

(2) $(a^n)^m = a^{mn}$.

(3) $a^m a^n = a^{m+n}$.

证明 留给读者作为练习.

Definition 2.1.16 Let G be a group and let $a \in G$. If $a^k = e$ for some $k \geqslant 1$, then the smallest such exponent $k \geqslant 1$ is called the **order** (阶) of a; if no such power exists, then one says that a has **infinite order**(无限阶).

We can see every element in a finite group has finite order. In any group G, the identity has order 1, and it is the only element in G of order 1; an element has order 2 if and only if it is not the identity and it is equal to its own inverse.

Lemma 2.1.17 Let G be a group and assume that $x \in G$ has finite order k. If $x^n = e$, then $k|n$.

证明 留给读者自证.

At last, we give a notion of a **dihedral group**(二面体群).

Definition 2.1.18 If π_n is a regular polygon with $n(\geqslant 3)$ vertices $v_1, v_2, \cdots, v_n$ and center O, then the symmetry group $\sum(\pi_n)$ is called the dihedral group of order $2n$, and it is denoted by D_{2n}. We define the dihedral group $D_4 = V$, the **four-group**, to be the group of order 4

$$V = \{(1), (1\ 2)(3\ 4), (1\ 3)(2\ 4), (1\ 4)(2\ 3)\} \subseteq S_4.$$

Remark 2.1.19 Some authors define the dihedral group D_{2n} as a group of order $2n$ generated by elements a, b such that $a^n = 1$, $b^2 = 1$ and $bab = a^{-1}$. Of course, one is obliged to prove existence of such a group.

2.2 子群的结构 (Structures of Subgroups)

本节主要介绍子群 (subgroup) 及其结构.

Definition 2.2.1 Let $*$ be an operation on a set G, and let $S \subseteq G$ be a subset. We say that S is **closed** under $*$ if $x * y \in S$ for all $x, y \in S$.

Recall that the operation on a group G is a function $* : G \times G \to G$. If $S \subseteq G$, then $S \times S \subseteq G \times G$, and to say that S is closed under the operation $*$ means that $*(S \times S) \subseteq S$. For example, the subset $\mathbb{Z}$ of the additive group $\mathbb{Q}$ of rational numbers is closed under $+$. However, if $\mathbb{Q}^\times$ is the multiplicative group of nonzero rational numbers, then $\mathbb{Q}^\times$ is closed under multiplication, but it is not closed under $+$ (for example, 2 and -2 lie in $\mathbb{Q}^\times$, but their sum $-2 + 2 = 0 \notin \mathbb{Q}^\times$).

Definition 2.2.2 A subset H of a group G is a **subgroup** (子群) if the following conditions (1)~(3) hold:

(1) $1 \in H$;

(2) if $x, y \in H$, then $xy \in H$, that is, H is closed under $*$;

(3) if $x \in H$, then $x^{-1} \in H$.

We write $H \leqslant G$ to denote H being a subgroup of a group G. Observe that $\{1\}$ and G are always subgroups of a group G, where $\{1\}$ denotes the subset consisting of the single element 1. We call a subgroup H of G **proper** (真子群) if $H \neq G$, and we write $H < G$. We call a subgroup H of G **nontrivial** (非平凡) if $H \neq \{1\}$. More interesting examples of subgroups will be given below.

Proposition 2.2.3 Every subgroup $H \leqslant G$ is itself a group.

证明 根据子群定义中条件 (2), H 在群运算下是封闭的, 也就是说, H 中有一个二元运算 (即 $* : G \times G \to G$ 在 $H \times H$ 上的限制). 这个运算是结合的 (associative): 由于对所有的 $x, y, z \in G, (xy)z = x(yz)$ 成立, 那么对所有的 $x, y, z \in H, (xy)z = x(yz)$ 亦成立. 最后, 由子群定义条件 (1) 和条件 (3) 可知, H 中含有单位元 (identity) 和逆元 (inverse). □

It is quicker to check that a subset H of a group G is a subgroup (and hence it is a group in its own right) than to verify the group axioms for H, for associativity is inherited from the operation on G and hence it needs not to be verified again.

Example 2.2.4 (1) The four permutations

$$V = \{(1), (1\ 2)(3\ 4), (1\ 3)(2\ 4), (1\ 4)(2\ 3)\}$$

form a group(dihedral group), because V is a subgroup of S_4: $(1) \in V$; $\alpha^2 = (1)$ for each $\alpha \in V$ and so $\alpha^{-1} = \alpha \in V$; the product of any two distinct permutations in $V - \{(1)\}$ is the third one. One calls V the **four-group** (or the **Klein group**) (V abbreviates the original German term Vierergruppe). Consider what verifying associativity $a(bc) = (ab)c$ would involve: there are 4 choices for each of a, b and c, and so there are $4^3 = 64$ equations to be checked. Of course, we may assume that none is (1), leaving us with only $3^3 = 27$ equations, but obviously, proving V is a group by showing it is a subgroup of S_4 is the best way to proceed.

(2) If $\mathbb{R}^2$ is the plane considered as an (additive) Abelian group, then any line L through the origin is a subgroup. The easiest way to see this is to choose a nonzero point (a, b) on L and then note that L consists of all the scalar multiples (ra, rb). The reader may now verify that the axioms in the definition of subgroup do hold for L.

Proposition 2.2.5 A subset H of a group G is a subgroup if and only if H is nonempty, and whenever $x, y \in H$, then $xy^{-1} \in H$.

证明 如果 H 是一个子群, 则它一定非空, 且 $1 \in H$. 如果 $x, y \in H$, 由子群定义条件 (3), 有 $y^{-1} \in H$, 由子群定义的条件 (2), 有 $xy^{-1} \in H$.

反过来, 假设 H 是满足上述条件的子集. 由于 H 是非空的, 不妨设 $h \in H$. 取 $x = h = y$, 则有 $1 = hh^{-1} \in H$, 这样就满足子群定义中条件 (1). 如果 $y \in H$, 取 $x = 1$ (这是合理的, 因为已证得 $1 \in H$), 可得 $y^{-1} = 1y^{-1} \in H$, 条件 (3) 满足. 最后, 我们知道 $(y^{-1})^{-1} = y$, 如果 $x, y \in H$, 则 $y^{-1} \in H$, 所以 $xy = x(y^{-1})^{-1} \in H$.

综上, H 是 G 的子群. □

Since every subgroup contains 1, one may replace the hypothesis "H is nonempty" by "$1 \in H$."

Note that if the operation in G is addition, then the condition in the proposition is that H is nonempty subset such that $x, y \in H$ implies $x - y \in H$.

For Galois, a group was just a subset H of S_n which is closed under composition, that is, if $\alpha, \beta \in H$, then $\alpha\beta \in H$. A. Cayley, in 1854, was the first to define an abstract group, mentioning associativity, inverses, and identity explicitly.

Proposition 2.2.6 A nonempty subset H of a finite group G is a subgroup if and only if H is closed under the operation of G, that is, if $a, b \in H$, then $ab \in H$. In particular, a nonempty subset of S_n is a subgroup if and only if it is closed under composition.

证明 留给读者自证.

This last proposition can be false when G is an infinite group. For example, the subset $\mathbb{N}$ of the additive group $\mathbb{Z}$ is closed under $+$, but it is not a subgroup of $\mathbb{Z}$(think why).

Definition 2.2.7 If G is a finite group, then the number $|G|$ of elements in G is called the **order** (阶) of G.

The word **order** has two different meanings in group theory: the order of an element $a \in G$ and the order $|G|$ of a group G. In the following section we will show for a cyclic group, the order of a generator is equal to the order of group itself.

Here is a way to constructing a new subgroup from given ones.

Proposition 2.2.8 The intersection $\cap_{i \in I} H_i$ of any family of subgroups of a group G is again a subgroup of G. In particular, if H and K are subgroups of G, then $H \cap K$ is a subgroup of G.

证明 设 $D = \cap_{i \in I} H_i$, 我们利用子群定义来证明 D 是 G 的子群. 首先 $D \neq \varnothing$, 因为 $1 \in H_i, \forall i$, 所以 $1 \in D$. 如果 $x \in D$, 则 x 在每一个 H_i 中; 由于每一个 H_i 都是子群, 则 $x^{-1} \in H_i, \forall i$, 即 $x^{-1} \in D$. 最后, 如果 $x, y \in D$, 则 x 和 y 都属于每一个 H_i, 因此乘积 xy 在每一个 H_i 中, 所以 $xy \in D$. □

Perhaps the most fundamental fact about subgroups H of a finite group G is that their orders are constrained. Certainly, we have $|H| \leqslant |G|$, but it turns out that

$|H|$ must be a divisor of $|G|$. To prove this, we introduce the notion of coset.

Definition 2.2.9 If H is a subgroup of a group G and $a \in G$, then the **coset**(陪集) aH is the subset of G, where

$$aH = \{ah : h \in H\}.$$

Of course, $a = a1 \in aH$. Cosets are usually not subgroups. For example, if $a \notin H$, then $1 \notin aH$ (otherwise $1 = ah$ for some $h \in H$, and this gives the contradiction $a = h^{-1} \in H$).

If we use the $*$ notation for the operation in a group G, then we denote the coset aH by $a * H$, where

$$a * H = \{a * h : h \in H\}.$$

In particular, if the operation is addition, then the coset is denoted by

$$a + H = \{a + h : h \in H\}.$$

If H is a subgroup of a group G, then the relation on G, defined by

$$a \equiv b, \quad \text{if} \quad a^{-1}b \in H,$$

is an equivalence relation on G. If $a \in G$, then $a^{-1}a = 1 \in H$, and $a \equiv a$; hence, $\equiv$ is reflexive. If $a \equiv b$, then $a^{-1}b \in H$; since subgroups are closed under inverses, $(a^{-1}b)^{-1} = b^{-1}a \in H$ and $b \equiv a$; hence $\equiv$ is symmetric. If $a \equiv b$ and $b \equiv c$, then $a^{-1}b, b^{-1}c \in H$; since subgroups are closed under multiplication, $(a^{-1}b)(b^{-1}c) = a^{-1}c \in H$, and $a \equiv c$. Therefore, $\equiv$ is transitive, and hence it is an equivalence relation.

We claim that the equivalence class of $a \in H$ is the coset aH. If $x \equiv a$, then there is $h \in H$ with $a^{-1}x = h$; hence, $x = ah \in aH$, and $\bar{a} \subseteq aH$. For the reverse inclusion, it is easy to see that if $x = ah \in aH$, then $x^{-1}a = (ah)^{-1}a = h^{-1}a^{-1}a = h^{-1} \in H$, so that $x \equiv a$ and $x \in \bar{a}$. Hence, $aH \subseteq \bar{a}$, and so $\bar{a} = aH$.

Lemma 2.2.10 Let H be a subgroup of a group G, and let $a, b \in G$.

(1) $aH = bH$ if and only if $b^{-1}a \in H$. In particular, $aH = H$ if and only if $a \in H$.

(2) If $aH \cap bH \neq \varnothing$, then $aH = bH$.

(3) $|aH| = |H|$ for all $a \in G$.

证明　留给读者作为练习.

Theorem 2.2.11 (Lagrange's Theorem)　If H is a subgroup of a finite group G, then $|H|$ is a divisor of $|G|$.

证明　设 $\{a_1H, a_2H, \cdots, a_tH\}$ 是所有 H 在 G 中不同的陪集. 因为陪集是等价类, 所以, 由定理 1.2.8 我们有

$$G = a_1H \cup a_2H \cup \cdots \cup a_tH,$$

故

$$|G| = |a_1H| + |a_2H| + \cdots + |a_tH|.$$

而对每个 i, $|a_iH| = |H|$, 所以 $|G| = t|H|$. □

Definition 2.2.12　The **index(指数)** of a subgroup H in G, denoted by $[G : H]$, is the number of cosets of H in G.

When G is finite, the index $[G : H]$ is the number t in the formula $|G| = t|H|$ in the proof of Lagrange's theorem, so that

$$|G| = [G : H]|H|.$$

This formula shows that the index $[G : H]$ is also a divisor of $|G|$.

Corollary 2.2.13　If H is a subgroup of a finite group G, then

$$[G : H] = |G|/|H|.$$

证明　留给读者作为练习.

Corollary 2.2.14　If G is a finite group and $a \in G$, then the order of a divides $|G|$.

证明　留给读者作为练习.

Corollary 2.2.15　If a finite group G has order m, then $a^m = 1$ for all $a \in G$.

证明　由推论 2.2.14, a 的阶为 d, 其中 $d|m$, 即存在整数 k 使得 $m = dk$. 从而, $a^m = a^{dk} = (a^d)^k = 1$. □

2.3 群同态 (Homomorphisms)

An important problem is to determine whether two given groups G and H are somehow the same. For example, we have investigated S_3, the group of all permutations of $X = \{1, 2, 3\}$. The group S_Y of all the permutations of $Y = \{a, b, c\}$ is a group different from S_3 because permutations of $\{1, 2, 3\}$ are different from those of $\{a, b, c\}$. Even though S_3 and S_Y are different, they surely bear a striking resemblance to each other. The notions of homomorphism and isomorphism allow one to compare different groups, as we shall see.

Definition 2.3.1 If $(G, *)$ and $(H, \circ)$ are groups (we have displayed the operation in each one), then a function $f : G \to H$ is a **homomorphism** (同态) if

$$f(x * y) = f(x) \circ f(y)$$

for all $x, y \in G$. If f is also a bijection, then f is called an **isomorphism**(同构). We say that G and H are **isomorphic**, denoted by $G \cong H$, if there exists an isomorphism $f : G \to H$. If $G = H$, then we say that f is an **endomorphism**(自同态) of G. Similar to **automorphism** (自同构).

Two obvious examples of homomorphisms are the identity $1_G : G \to G$, which is an isomorphism, and the trivial homomorphism $f : G \to H$, defined by $f(a) = 1$ for all $a \in G$.

Here are more interesting examples. Let $\mathbb{R}$ be the group of all real numbers with operation addition, and let $\mathbb{R}^>$ be the group of all positive real numbers with operation multiplication. The function $f : \mathbb{R} \to \mathbb{R}^>$, defined by $f(x) = e^x$, is a homomorphism, for if $x, y \in \mathbb{R}$, then

$$f(x + y) = e^{x+y} = e^x e^y = f(x)f(y).$$

Now f is an isomorphism, for its inverse function $g : \mathbb{R}^> \to \mathbb{R}$ is defined by $\log(x)$. Therefore, the additive group $\mathbb{R}$ is isomorphic to the multiplicative group $\mathbb{R}^>$. Note that the inverse function g is also an isomorphism:

$$g(xy) = \log(xy) = \log(x) + \log(y) = g(x) + g(y).$$

Lemma 2.3.2 Let $f : G \to H$ be a homomorphism. Then

(1) $f(1) = 1$;

(2) $f(x^{-1}) = f(x)^{-1}$;

(3) $f(x^n) = f(x)^n$ for all $n \in \mathbb{Z}$.

证明　(1) 用 f 作用于 G 中公式 $1 \cdot 1 = 1$, 有 $f(1)f(1) = f(1)$, 其中 $f(1) \in H$, 在此式两边都乘以 $f(1)^{-1}$, 有 $f(1) = 1$.

(2) 用 f 作用于 G 中公式 $x^{-1}x = 1$, 有 $f(x^{-1})f(x) = 1$, 由逆元的唯一性, $f(x^{-1}) = f(x)^{-1}$.

(3) 由定义易得对所有的 $n \geqslant 0$, 有 $f(x^n) = f(x)^n$. 对于负数 n, 有 $(y^{-1})^n = y^{-n}$, 对群中所有元素 y, 因此

$$f(x^{-n}) = f((x^{-1})^n) = (f(x^{-1}))^n = (f(x)^{-1})^n = f(x)^{-n}. \qquad \square$$

Definition 2.3.3　If $f : G \to H$ is a homomorphism, define

$$\textbf{kernel } f = \{x \in G : f(x) = 1\}$$

and

$$\textbf{image } f = \{h \in H : h = f(x) \text{ for some } x \in G\}.$$

We usually abbreviate kernel f to ker f and image f to im f.

Proposition 2.3.4　Let $f : G \to H$ be a homomorphism. Then

(1) ker f is a subgroup of G and im f is a subgroup of H;

(2) if $x \in \ker f$ and if $a \in G$, then $axa^{-1} \in \ker f$;

(3) f is an injection if and only if $\ker f = \{1\}$.

证明　(1) 由引理 2.3.2 知, $f(1) = 1$, 故 $1 \in \ker f$. 如果 $x, y \in \ker f$, 则 $f(x) = 1 = f(y)$, 因此, $f(xy) = f(x)f(y) = 1 \cdot 1 = 1$, 所以 $xy \in \ker f$. 最后, 如果 $x \in \ker f$, 则 $f(x) = 1$, 所以 $f(x^{-1}) = f(x)^{-1} = 1^{-1} = 1$, 即 $x^{-1} \in \ker f$. 因此, ker f 是 G 的子群.

我们现在证明 im f 是 H 的子群. 首先, $1 = f(1) \in \operatorname{im} f$. 接下来, 如果 $h = f(x) \in \operatorname{im} f$, 则 $h^{-1} = f(x)^{-1} = f(x^{-1}) \in \operatorname{im} f$. 最后, 如果 $k = f(y) \in \operatorname{im} f$, 则 $hk = f(x)f(y) = f(xy) \in \operatorname{im} f$. 因此, im f 是 H 的子群.

(2) 如果 $x \in \ker f$, 则 $f(x) = 1$ 且

$$f(axa^{-1}) = f(a)f(x)f(a)^{-1} = f(a)1f(a)^{-1} = f(a)f(a)^{-1} = 1.$$

因此, $axa^{-1} \in \ker f$.

(3) 如果 f 是单射, 则 $x \neq 1$, 那么 $f(x) \neq f(1) = 1$, 所以 $x \notin \ker f$. 反过来, 假设 $\ker f = \{1\}$, $f(x) = f(y)$, 那么, $1 = f(x)f(y)^{-1} = f(xy^{-1})$, 于是 $xy^{-1} \in \ker f = 1$. 因此, $xy^{-1} = 1$, 即 $x = y$, 故 f 是单射. □

Definition 2.3.5 A subgroup K of a group G is called a **normal subgroup** (正规子群), if $k \in K$ and $g \in G$ imply $gkg^{-1} \in K$. If K is a normal subgroup of G, one writes

$$K \lhd G.$$

Example 2.3.6 (1) Define the **center**(中心) of a group G, denoted by $Z(G)$, to be

$$Z(G) = \{z \in G : zg = gz \text{ for all } g \in G\},$$

that is, $Z(G)$ consists of all elements commuting with everything in G.

It is easy to check that $Z(G)$ is a subgroup of G, and this work is left to the reader.

The center $Z(G)$ is a normal subgroup: if $z \in Z(G)$ and $g \in G$, then

$$gzg^{-1} = zgg^{-1} = z \in Z(G).$$

(2) We can see from the proposition that the kernel of a homomorphism is always a normal subgroup.

(3) If G is an Abelian group, then every subgroup K is normal, for if $k \in K$ and $g \in G$, then $gkg^{-1} = kgg^{-1} = k \in K$.

Definition 2.3.7 If G is a group and $a \in G$, then a **conjugate** (共轭元) of a is an element in G of the form

$$gag^{-1},$$

where $g \in G$.

It is clear that a subgroup $K \leqslant G$ is a normal subgroup if and only if K contains all the conjugates of its elements: if $k \in K$, then $gkg^{-1} \in K$ for all $g \in G$.

Definition 2.3.8 If G is a group and $g \in G$, define **conjugation** (共轭) $\gamma_g : G \to G$ by

$$\gamma_g(a) = gag^{-1}$$

for all $a \in G$.

Proposition 2.3.9 (1) If G is a group and $g \in G$, then conjugation $\gamma_g : G \to G$ is an isomorphism.

(2) Conjugate elements have the same order.

证明 留给读者作为练习.

Theorem 2.3.10 (Cayley) Every group G is (isomorphic to) a subgroup of the symmetric group S_G. In particular, if $|G| = n$, then G is isomorphic to a subgroup of S_n.

证明 对每个 $a \in G$, 定义"平移" $\tau_a : G \to G$: 对每个 $x \in G$ 有 $\tau_a(x) = ax$ (若 $a \neq 1$, 则 τ_α 不是一个同态). 对 $a, b \in G$, $(\tau_a \circ \tau_b)(x) = \tau_a(bx) = a(bx) = (ab)x = \tau_{ab}(x)$, 所以

$$\tau_a \tau_b = \tau_{ab}.$$

于是每个 τ_a 是双射, 因它的逆是 $\tau_{a^{-1}}$:

$$\tau_a \tau_{a^{-1}} = \tau_{aa^{-1}} = \tau_1 = 1_G,$$

所以 $\tau_a \in S_G$.

定义 $\varphi : G \to S_G$, $\varphi(a) = \tau_a$, 则

$$\varphi(a)\varphi(b) = \tau_a \tau_b = \tau_{ab} = \varphi(ab),$$

所以 φ 是一个同态. 最后, φ 是一个单射. 若 $\varphi(a) = \varphi(b)$, $\tau_a = \tau_b$, 因此对所有 $x \in G$ 有 $\tau_a(x) = \tau_b(x)$. 特别地, 当 $x = 1$ 时, 得 $a = b$.

后一个命题的证明留给读者, 读者可证: 当 X 是集合且 $|X| = n$, 则 $S_X \cong S_n$. □

2.4 循环群 (Cyclic Groups)

本节给出了循环群 (cyclic group) 的定义与结构定理.

Definition 2.4.1 (1) If G is a group and $a \in G$, write

$$\langle a \rangle = \{a^n | n \in \mathbb{Z}\} = \{\text{ all powers of } a\};$$

$\langle a \rangle$ is called the **cyclic subgroup** of G **generated** by a (由 a 生成的 G 的循环群).

(2) A group G is called **cyclic** if there is some $a \in G$ with $G = \langle a \rangle$; in this case, a is called a **generator** (生成子) of G.

It is easy to see that $\langle a\rangle$ is a subgroup. A cyclic group can have several different generators. For example, $\langle a\rangle=\langle a^{-1}\rangle$.

Example 2.4.2 (1) The additive group of integers $\mathbb{Z}$ is cyclic with the generator 1.

(2) For every positive integer n, $\mathbb{Z}_n=\{\overline{0},\overline{1},\cdots,\overline{n-1}\}$ (see Theorem 1.2.4) is a cyclic group under the operation defined by

$$\overline{a}+\overline{b}=\overline{a+b}$$

for all $\overline{a},\overline{b}\in\mathbb{Z}_n$. The generator is $\overline{1}$.

Proposition 2.4.3 If $G=\langle a\rangle$ is a cyclic group of order n, then a^k is a generator of G if and only if $(k,n)=1$.

证明 如果 a^k 是一个生成子, 则 $a\in\langle a^k\rangle$, 所以存在 s 使得 $a=a^{ks}$. 因此, $a^{ks-1}=1$, 所以 $n|(ks-1)$. 也就是说, 存在整数 t 使得 $ks-1=tn$, 即 $sk-tn=1$. 因此, $(k,n)=1$.

反过来, 由于 $(k,n)=1$, 存在整数 s 和 t 使得 $1=sk+tn$. 因此, $a=a^{sk+tn}=a^{sk}$ (因为 $a^{tn}=1$), 所以 $a\in\langle a^k\rangle$. 于是, $G=\langle a\rangle\leqslant\langle a^k\rangle$, 所以 $G=\langle a^k\rangle$. □

Corollary 2.4.4 The number of generators of a cyclic group of order n is $\phi(n)$, where ϕ is the Euler function, i.e. $\phi(n)=n(1-1/p_1)(1-1/p_2)\cdots(1-1/p_k)$ if the prime factorization of n is given by $n=p_1^{a_1}\cdots p_k^{a_k}$.

Proposition 2.4.5 Every subgroup S of a cyclic group $G=\langle a\rangle$ is itself cyclic. In fact, a^k is a generator of S, where k is the smallest positive integer m with $a^m\in S$.

证明 我们可以假设 S 是非平凡的, 也就是说, $S\neq\{1\}$, 因为此命题在 $S=\{1\}$ 的情况下明显正确. 由假设, 存在 $0\neq n\in\mathbb{Z}$, 使得 $a^n\in H$. 于是 $a^{-n}\in H$, 从而

$$M=\{n\in\mathbb{N}|a^n\in H\}$$

是一个非空集合. 令 k 是 M 中最小的正整数. $\forall a^m\in H$, 设 $m=kq+t, 0\leqslant t<k$, 则 $a^t=a^{m-kq}=a^m(a^k)^{-q}\in H$. 由 k 的最小性知 $t=0$, 于是 $m=kq$. 所以 $a^m=a^{kq}=(a^k)^q$. 因此 $H=\langle a^k\rangle$. □

Proposition 2.4.6 Let G be a finite group and let $a\in G$. Then the order of a is the number of elements in $\langle a\rangle$.

证明 因为 G 是有限的, 所以存在整数 $k\geqslant 1$ 使得 $1,a,a^2,\cdots,a^{k-1}$ 是 G 中 k 个不同的元素, 而 $1,a,a^2,\cdots,a^k$ 出现重复. 因此 $a^k\in\{1,a,a^2,\cdots,a^{k-1}\}$, 即对

某个 $i, 0 \leqslant i < k$, 有 $a^k = a^i$. 若 $i \geqslant 1$, 则 $a^{k-i} = 1$, 此与 $1, a, a^2, \cdots, a^{k-1}$ 无重复矛盾. 因此, $a^k = a^0 = 1$, k 是 a 的阶 (因为 k 是满足阶条件的最小正整数).

设 $H = \{1, a, a^2, \cdots, a^{k-1}\}$, 则 $|H| = k$, 只需证明 $H = \langle a \rangle$. 显然, $H \subseteq \langle a \rangle$. 对于反包含, 取 $a^i \in \langle a \rangle$. 由除法算式知 $i = qk + r$, 其中 $0 \leqslant r < k$. 因此 $a^i = a^{qk+r} = a^{qk}a^r = (a^k)^q a^r = a^r \in H$, 由此得 $\langle a \rangle \subseteq H$. 所以 $H = \langle a \rangle$. □

Proposition 2.4.7 Let G be a group of order n. If G is cyclic, then G has a unique subgroup of order d for each divisor d of n. Conversely, if there is at most one cyclic subgroup of order d, where $d|n$, then G is cyclic.

证明 假设 $G = \langle a \rangle$ 是阶为 n 的循环群. 我们断言, $\langle a^{n/d} \rangle$ 的阶为 d. 显然, $(a^{n/d})^d = a^n = 1$, 所以只需证明 d 是满足阶条件的最小正整数. 若 $(a^{n/d})^r = 1$, 则 $n|(n/d)r$. 因而存在整数 s 满足 $(n/d)r = ns$, 则 $r = ds$, $r \geqslant d$.

为了证明唯一性, 设 C 是 G 的阶为 d 的子群, 那么 C 是循环群, 不妨设 $C = \langle x \rangle$. 现在 $x = a^m$ 的阶为 d, 所以 $1 = (x^m)^d$. 因而 $n|md$, 则 $md = nk$, k 为整数. 从而, $x = a^m = (a^{n/d})^k$, 所以 $C = \langle x \rangle \subseteq \langle a^{n/d} \rangle$. 又因为两个子群的阶相同, 所以 $C = \langle a^{n/d} \rangle$.

反之, 定义群 G 上的一个关系 $a \equiv b$, 若 $\langle a \rangle = \langle b \rangle$. 容易看出, 这是一个等价关系, 且 $a \in G$ 的等价类 $\bar{a}$ 由 $C = \langle a \rangle$ 的所有生成元构成. 因此, 我们用 $\mathrm{gen}(C)$ 表示 $\bar{a}$, 且

$$G = \bigcup_{C\text{是循环群}} \mathrm{gen}(C).$$

因而 $n = |G| = \sum_C |\mathrm{gen}(C)|$. 那么 $|\mathrm{gen}(C)| = \phi(|C|)$. 根据假设, G 至多有一个任意阶的循环子群, 所以

$$n = \sum_C |\mathrm{gen}(C)| \leqslant \sum_{d|n} \phi(d) = n,$$

因此, 对 n 的每个因子 d, 一定存在阶为 d 的循环子群 C, 分配 $\phi(d)$ 给 $\sum_C |\mathrm{gen}(C)|$. 特别地, 一定存在阶为 n 的循环子群 C, 所以 G 是循环群. □

Proposition 2.4.8 Let a and b be integers and let $A = \langle a \rangle$ and $b = \langle b \rangle$ be the cyclic subgroups of $\mathbb{Z}$ they generate.

(1) If $A + B$ is defined to be $\{sa + tb : s, t \in \mathbb{Z}\}$, then $A + B = \langle d \rangle$, where $d = (a, b)$.

(2) $A \cap B = \langle m \rangle$, where $m = [a, b]$.

证明 (1) 由命题 2.4.5, $A+B$ 是循环群, $A+B$ 是由 $A\cup B$ 生成的子群, 所以 $A+B=\langle d\rangle$, 其中 d 是 a 和 b 最小的线性组合, 即 $d=(a,b)$.

(2) 如果 $c\in A\cap B$, 则 $c\in A$, 即 $a|c$, 类似地, $c\in B$, $b|c$. 从而, $A\cap B$ 中的每个元素都是 a 和 b 的公倍数. 反之, a 和 b 的每一个公倍数都在 $A\cap B$ 中. 因为 $A\cap B$ 是 $\mathbb{Z}$ 的子群, 所以 $A\cap B$ 是循环群, $A\cap B=\langle m\rangle$, m 为 $A\cap B$ 中最小的非负数. 故 $m=[a,b]$. □

Proposition 2.4.9 If p is prime, then every group G of order p is cyclic.

证明 此命题可由 Lagrange 定理得到. 下面给出简单证明. 选择 $a\in G$, 其中 $a\neq 1$, 设 $H=\langle a\rangle$ 为 G 的循环子群. 由 Lagrange 定理, $|H|$ 是 $|G|=p$ 的因子. 由于 p 是素数, $|H|>1$, 因此 $|H|=|G|=p$, 故 $H=G$. □

In what follows, we will give the structure theorem of a cyclic group.

Theorem 2.4.10 Let G be a cyclic group with a generator a. Then

(1) if the order of a is infinite, then G is isomorphic to the additive group $\mathbb{Z}$;

(2) if the order of a is n, then G is isomorphic to the additive group $\mathbb{Z}_n$ given by Example 2.4.2(2).

证明 (1) 设 a 的阶无限. 此时,

$$a^h=a^k,\quad 当且仅当 h=k.$$

由 $h=k$, 可得 $a^h=a^k$, 显然. 而假如 $a^h=a^k$ 且 $h\neq k$, 我们可以假定 $h>k$, 从而得到 $a^{h-k}=e$, 这与 a 的阶无限的假定矛盾.

于是,

$$a^k\longrightarrow k$$

是 G 与整数加群 $\mathbb{Z}$ 间的一一映射, 且

$$a^ha^k=a^{h+k}\longrightarrow h+k$$

所以,

$$G\cong\mathbb{Z}.$$

(2) 如果 a 的阶是 n, $a^n=e$. 此时,

$$a^h=a^k,\quad 当且仅当 n|h-k.$$

假如 $a^h = a^k$, 那么 $h-k=nq$, $h=k+nq$,

$$a^h = a^{k+nq} = a^k(a^n)^q = a^k e^q = a^k$$

假如 $a^h = a^k$, 令 $h-k=nq+r$, $0 \leqslant r \leqslant n-1$, 那么

$$e = a^{h-k} = a^{nq+r} = a^{nq}a^r = ea^r = a^r.$$

由阶的定义, $r=0$, 也就是说, $n|h-k$.

于是

$$a^k \longrightarrow \bar{k}$$

是 G 与剩余类群 $\mathbb{Z}_n$ 间的一一映射, 且

$$a^h a^k = a^{h+k} \longrightarrow \overline{h+k} = \bar{h} + \bar{k},$$

故

$$G \cong \mathbb{Z}_n.$$

□

2.5 商群 (Quotient Groups)

本节将给出由已知群构造新群的方法, 这个新群就是商群 (**quotient group**).

Lemma 2.5.1 A subgroup K of a group G is a normal subgroup if and only if $bK = Kb$ for every $b \in G$.

证明 设 $bk \in bK$. 由于 K 是正规子群, 则 $bkb^{-1} \in K$. 记 $bkb^{-1} = k' \in K$, 于是 $bk = (bkb^{-1})b = k'b \in Kb$, 所以 $bK \subseteq Kb$. 再证反包含, 设 $kb \in Kb$. 由于 K 的正规性, $(b^{-1})k(b^{-1})^{-1} = b^{-1}kb = k \in K$. 记 $b^{-1}kb = k'' \in K$, 因此 $kb = b(b^{-1}kb) = bk'' \in bK$, 所以 $Kb \subseteq bK$. 于是, 当 $K \lhd G$ 时, $bK = Kb$.

下面证充分性, 假设 $bK = Kb$, $\forall b \in G$. 如果 $x \in K$, 则 $bx \in bK = Kb$, 因此, 存在 $x' \in K$ 使得 $bx = x'b$, 所以 $bxb^{-1} = x' \in K$. 故 $K \lhd G$. □

It follows from Lemma 2.5.1 that if $K \lhd G$, then every left coset of K in G is a right coset of K. Here is a fundamental construction of a new group from a given group.

Theorem 2.5.2 Let G/K denote the family of all the cosets of a subgroup K of G. If K is a normal subgroup, then

$$aKbK = abK$$

for all $a, b \in G$, and G/K is a group under this operation.

Remark 2.5.3 The group G/K is called the **quotient group** G mod K; when G is finite, its order $|G/K|$ is the index $[G : K] = |G|/|K|$ (I think this is possibly the reason quotient groups are so called).

证明 由引理 2.5.1 和结合性,

$$(aK)(bK) = a(Kb)K = a(bK)K = abKK = abK.$$

因为 K 是子群, 所以 $KK = K$, 从而两个 K 的陪集的乘积还是 K 的陪集. 故 G/K 上的运算已定义. 单位元是 $K = 1K$, 因为 $(1K)(bK) = 1bK = bK$; 且 aK 的逆是 $a^{-1}K$, 因为 $(a^{-1}K)(aK) = a^{-1}aK = K$. 故 G/K 是群. □

Example 2.5.4 We show that the quotient group $\mathbb{Z}/\langle m\rangle$ is precisely $\mathbb{Z}_m$, where $\langle m\rangle$ is the (cyclic) subgroup consisting of all the multiples of a positive integer m. Since $\mathbb{Z}$ is Abelian, $\langle m\rangle$ is necessarily a normal subgroup. The sets $\mathbb{Z}/\langle m\rangle$ and $\mathbb{Z}_m$ coincide because they are comprised of the same elements: the coset $a + \langle m\rangle$ is the congruence class $\overline{a}$, and here it is denoted by $[a]$, that is,

$$a + \langle m\rangle = \{a + km | k \in \mathbb{Z}\} = [a].$$

The operations also coincide: addition in $\mathbb{Z}/\langle m\rangle$ is given by

$$(a + \langle m\rangle) + (b + \langle m\rangle) = (a + b) + \langle m\rangle;$$

since $a + \langle m\rangle = [a]$, this last equation is just $[a] + [b] = [a + b]$, which is the sum in $\mathbb{Z}_m$ as defined by Example 2.4.2(2). Therefore, $\mathbb{Z}_m$ is equal to the quotient group $\mathbb{Z}/\langle m\rangle$.

Corollary 2.5.5 Every normal subgroup is the kernel of some homomorphism.

证明 如果 $K \lhd G$, 定义自然映射 (natural map)

$$\pi : G \to G/K, \quad a \mapsto aK.$$

上述定理中公式 $aKbK = abK$ 可以写成 $\pi(a)\pi(b) = \pi(ab)$, 由此, π 是一个满同态. 由于 K 是 G/K 中的单位元, 由引理 2.2.10(2),

$$\ker \pi = \{a \in G : \pi(a) = K\} = \{a \in G : aK = K\} = K.$$ □

2.6　群同态基本定理
(The Fundamental Theorem of Group Homomorphisms)

本节主要证明三个群同态基本定理.

The next theorem shows that every homomorphism gives rise to an isomorphism and that quotient groups are merely constructions of homomorphic images. Noether(诺特) emphasized the fundamental importance of this fact, and this theorem is often named after her.

Theorem 2.6.1　(The First Isomorphism Theorem) If $f : G \to H$ is a homomorphism, then

$$\ker f \lhd G \quad \text{and} \quad G/\ker f \cong \operatorname{im} f.$$

In more details, if $\ker f = K$, then $\varphi : G/K \to \operatorname{im} f \leqslant H$, given by $\varphi : aK \mapsto f(a)$, is an isomorphism.

Remark 2.6.2　The following diagram describes the proof of the First Isomorphism Theorem, where $\pi : G \to G/K$ is the natural map $a \mapsto aK$ and $i : \operatorname{im} f \to H$ is the inclusion.

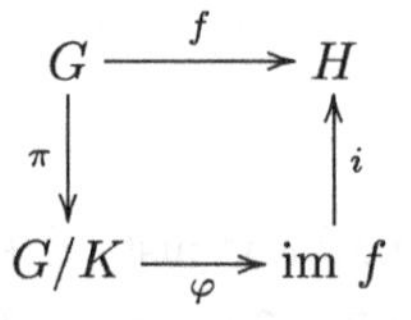

$$\begin{array}{ccc} G & \xrightarrow{f} & H \\ {\scriptstyle \pi}\downarrow & & \uparrow{\scriptstyle i} \\ G/K & \xrightarrow[\varphi]{} & \operatorname{im} f \end{array}$$

证明　由命题 2.3.4 知, $K = \ker f$ 是 G 的正规子群. φ 是良定义的 (well-defined): 如果 $aK = bK$, 则有某个 $k \in K$ 使得 $a = bk$, 因此 $f(a) = f(bk) = f(b)f(k) = f(b)$, 由于 $f(k) = 1$.

现在证明 φ 是同态. 由于 f 是同态且 $\varphi(aK) = f(a)$,

$$\varphi(aKbK) = \varphi(abK) = f(ab) = f(a)f(b) = \varphi(aK)\varphi(bK).$$

易见 $\operatorname{im} \varphi \leqslant \operatorname{im} f$. 对于反包含, 注意到 $y \in \operatorname{im} f$, 则存在 $a \in G$, 使得 $y = f(a)$, 从而, $y = f(a) = \varphi(aK)$. 从而, φ 是满射.

最后, 证明 φ 是单射. 如果 $\varphi(aK) = \varphi(bK)$, 则 $f(a) = f(b)$. 因此, $1 = f(b)^{-1}f(a) = f(b^{-1}a)$, 即 $b^{-1}a \in \ker f = K$. 由此, $aK = bK$, 所以 φ 是单射. 故 $\varphi : G/K \to \operatorname{im} f$ 是同构. $\square$

Note that $i\varphi\pi = f$, where $\pi : G \to G/K$ is the natural map and $i : \text{im } f \to H$ is the inclusion, so that f can be reconstructed from G and $K = \ker f$.

Given any homomorphism $f : G \to H$, we should immediately ask for its kernel and image; the First Isomorphism Theorem will then provide an isomorphism $G/\ker f \cong \text{im } f$. Since there is no significant difference between isomorphic groups, the First Isomorphism Theorem also indicates that there is no significant difference between quotient groups and homomorphic images.

Example 2.6.3 What is the quotient group $\mathbb{R}/\mathbb{Z}$? Take the real line and identify integer points, which amounts to taking the unit interval $[0, 1]$ and identifying its endpoints, yielding the circle. Define $f : \mathbb{R} \to S^1$, where S^1 is the circle group, by

$$f : x \mapsto e^{2\pi i x}.$$

Now f is a homomorphism, that is, $f(x + y) = f(x)f(y)$. The map f is surjective and $\ker f$ consists of all $x \in \mathbb{R}$ for which $e^{2\pi i x} = \cos 2\pi x + i \sin 2\pi x = 1$, that is, $\cos 2\pi x = 1$ and $\sin 2\pi x = 0$. But $\cos 2\pi x = 1$ forces x to be an integer; since $1 \in \ker f$, we have $\ker f = \mathbb{Z}$. The First Isomorphism Theorem now gives

$$\mathbb{R}/\mathbb{Z} \cong S^1.$$

A natural question is whether HK is a subgroup when both H and K are subgroups. In general, HK needs not be a subgroup. For example, let $G = S_3$, $H = \langle(1\ 2)\rangle$, and $K = \langle(1\ 3)\rangle$. Then

$$HK = \{(1), (1\ 2), (1\ 3), (1\ 3\ 2)\}$$

is not a subgroup lest we contradict Lagrange's theorem. But we left a question to the reader, whether there is a necessary and sufficient condition for the product HK of subgroups H and K to be a subgroup. The answer is positive and the reader may prove it in the exercises of this chapter.

Here is a counting result.

Proposition 2.6.4 If H and K are subgroups of a finite group G, then

$$|HK||H \cap K| = |H||K|.$$

Remark 2.6.5 The subset $HK = \{hk | h \in H \text{ and } k \in K\}$ needs not be a subgroup of G.

证明 定义函数 $f: H\times K\to HK$, $f:(h,k)\mapsto hk$. 显然, f 是满射. 只需证明: 对每个 $x\in HK$ 有 $|f^{-1}(x)|=|H\cap K|$, 其中 $f^{-1}(x)=\{(h,k)\in H\times K: f(h,k)=x\}$.

我们断言, 若 $x=hk$, 则

$$f^{-1}(x)=\{(hd,d^{-1}k):d\in H\cap k\}.$$

因为 $f(hd,d^{-1}k)=hdd^{-1}k=hk=x$, 所以 $(hd,d^{-1}k)\in f^{-1}(x)$. 对于反包含, 设 $(h',k')\in f^{-1}(x)$, 则 $h'k'=hk$. 从而 $h^{-1}h'=kk'^{-1}\in H\cap K$, 称这个元素为 d, 则 $h'=hd$, $k'=d^{-1}k$, 所以 (h',k') 属于等式右边. 故

$$|f^{-1}(x)|=|\{(hd,d^{-1}k):d\in H\cap K\}|=|H\cap K|,$$

由于 $d\mapsto(hd,d^{-1}k)$ 是一个双射. □

The next two results are consequences of the First Isomorphism Theorem.

Theorem 2.6.6 (The Second Isomorphism Theorem) If H and K are subgroups of a group G with $H\lhd G$, then HK is a subgroup, $H\cap K\lhd K$, and

$$K/(H\cap K)\cong HK/H.$$

证明 先证明 HK/H 是有意义的, 然后再描述其元素. 由于 $H\lhd G$, 而 HK 是子群, 从一个更一般的事实中可得 H 在 HK 中的正规性: 若 $H\leqslant S\leqslant G$ 且 H 在 G 中正规, 则 H 在 S 中正规 (若对每个 $g\in G$ 有 $ghg^{-1}\in H$, 则对每个 $g\in S$ 有 $ghg^{-1}\in H$).

现在证明每个陪集 $xH\in HK/H$ 有形式 $kH, k\in K$. 当然, $xH=hkH$, 其中 $h\in H$, $k\in K$. 但是对某个 $h'\in H$ 有 $hk=k(k^{-1}hk)=kh'$, 所以 $hkH=kh'H=kH$.

于是函数 $f:K\to HK/H(k\mapsto kH)$ 是满射. 另外, f 是一个同态, 因为它是自然映射 $\pi:G\to G/H$ 限制在 K 上. 由于 $\ker\pi=H$, 所以 $\ker f=H\cap K$, 故 $H\cap K$ 是 K 的正规子群. 由第一同构定理知 $K/(H\cap K)\cong HK/H$. □

The Second Isomorphism Theorem gives the product formula in the special case when one of the subgroups is normal: if $K/(H\cap K)\cong HK/H$, then $|K/(H\cap K)|=|HK/H|$, and so $|HK||H\cap K|=|H||K|$.

Theorem 2.6.7 (The Third Isomorphism Theorem) If H and K are normal subgroups of a group G with $K\leqslant H$, then $H/K\lhd G/K$ and

$$(G/K)/(H/K)\cong G/H.$$

证明 定义 $f: G/K \to G/H, aK \mapsto aH$. 注意到 f 是一个良定义的函数, 因为若 $a' \in G$ 和 $a'K = aK$, 则 $a^{-1}a' \in K \leqslant H$, 所以 $aH = a'H$. 容易看出 f 是满同态.

因为 $f(aK) = H$ 当且仅当 $a \in H$, 所以 $\ker f = H/K$, 则 H/K 是 G/K 的正规子群. 由于 f 是满射, 因此由第一同构定理知 $(G/K)/(H/K) \cong G/H$. □

Remark 2.6.8 The third isomorphism theorem is easy to remember: the K's in the fraction $(G/K)/(H/K)$ can be canceled. One can better appreciate the first isomorphism theorem after having proved the third one. The elements of $(G/K)/(H/K)$ are cosets of H/K whose representatives are themselves cosets (of G/K). A direct proof of the third isomorphism theorem could be nasty.

2.7 应用 (Applications)

Group theory is an indispensable mathematical tool in many branches of chemistry and physics. This section provides a self-contained and rigorous account on the fundamentals and applications of the subject to chemical physics.

Quantum mechanics showed that the elementary systems that matter is made of, such as electrons and protons, are truly identical, not just very similar, so that symmetry in their arrangement is exact, not approximate as in the macroscopic world. Systems were also seen to be described by functions of position that are subject to the usual symmetry operations of rotation and reflection, as well as to others not so easily described in concrete terms, such as the exchange of identical particles. Elementary particles were observed to reflect symmetry properties in more esoteric spaces. In all these cases, symmetry can be expressed by certain operations on the systems concerned, which have properties revealed by group theory, a rather obscure branch of mathematics that had previously been mainly a curiosity without practical application.

Physics uses that part of group theory known as the theory of representations, in which matrices acting on the members of a vector space is the central theme. It allows certain members of the space to be created that are symmetrical, and which can be classified by their symmetry. It is found that all the observed spectroscopic states of atoms and molecules correspond to such symmetrical functions, and can be classified accordingly. Among other things, it gives selection rules that specify which transitions are observed, and which are not. These matters are so commonplace in spectroscopy that the fact that they are extraordinary and wonderful is hardly realized.

The Hamiltonian operator H is a function of the coordinates and momenta of a system. In the Schrödinger representation, solutions of the eigenvalue equation $H\psi = E\psi$ exist for certain values of the number E that can be cast into an orthonormal set of functions of the coordinates $\{\psi\}$, the eigenfunctions. If more than one state ψ corresponds to the same energy E, this energy level is said to be degenerate (not the states, since there is nothing degenerate about the states involved–they simply have the same energy). In general, this does not occur except when there is some reason forcing the states to have the same energy, as we shall see. When it occurs without a compelling reason, it is called accidental degeneracy.

Multiply the eigenvalue equation by some operator Q from the left, and insert $Q^{-1}Q$ between H and ψ, to get $(QHQ^{-1})Q\psi = QE\psi = EQ\psi$. If $QHQ^{-1} = H$, or what is the same thing, $QH = HQ$, then $HQ\psi = EQ\psi$, or the states ψ and $Q\psi$ belong to the same value of the energy E; that is, the level is degenerate. If Q represents some change of basis functions or coordinates, QHQ^{-1} is the operator H in the new frame of reference. Should Q be a symmetry operation, then we must have $QHQ^{-1} = H$. The set of operators commuting with H is a group, called the symmetry group of the Hamiltonian.

It is often useful to express H as the sum of H_0, an approximate Hamiltonian that is simple and corresponds to most of the energy, and a perturbation Hamiltonian H' that includes finer details. If the symmetry group of H' is smaller than the symmetry group of H_0, then the greater symmetry is broken by the perturbation. In this case, previously degenerate levels may split into distinct levels, some of which may still be degenerate. Group theory will provide suitable functions for this calculation that can greatly reduce the effort involved.

Example 2.7.1 The states of an isolated atom are classified by the total angular momentum J, and belong to irreducible representations of the rotation group in three dimensions with dimension $2J + 1$. As long as there is spherical symmetry, these states all have exactly the same energy. If you apply a magnetic field, the spherical symmetry is broken, and there is now only symmetry about an axis in the direction of the magnetic field. The $2J + 1$ levels now acquire different energies, and the nondegenerate states can be classified by the magnetic quantum number M, $M = J, J - 1, \cdots, -J$. Group theory can determine these states in advance, so that the splitting is given by a simple diagonal matrix element.

Example 2.7.2 The energy levels of the hydrogen atom depend only on the principal quantum number N. For any N, there are states with $J = 0, 1, \cdots, N-1$. As far as the rotation group is concerned, this degeneracy is accidental. In fact, the Hamiltonian is invariant under the four-dimensional rotation group (actually, a group isomorphic to the four-dimensional rotation group), and its irreducible representations explain the added degeneracy, which is really not accidental at all. The degeneracy is lifted when the potential has a different radial dependence in more complex atoms, although the spherical symmetry is still there.

Example 2.7.3 It was noted in the late 1960's that the nucleons (neutron and proton) and the Lambda were not greatly different in mass. It was conjectured that they were members of an irreducible representation of dimension 3 of a group called SU_3, and that the differences in the observed masses were a splitting due to a perturbation related to the property called strangeness. There was some success in arranging the known particles in irreducible representations of SU_3, but later a theory of the internal structure of heavy elementary particles superseded this idea. However, group theory is still quite important in the field.

Group theory is good for more than classification, however. The probability amplitude for a transition between two states ψ and φ is given by some integral like $\int \psi^* P\varphi d\tau$, where P is an operator characteristic of the mechanism involved in the transition (electric dipole, etc), and $d\tau$ is the differential volume element. If ψ, φ and P are all classifed according to their irreducible representations, then their product belongs to a representation whose characters are the product of the characters of each of the representations involved. This representation is almost always reducible, and the component irreducible representations can be found by character analysis. Unless the unit representation (all matrices $+1$) is there, the integral must vanish, since otherwise it changes under some symmetry operation, which by definition cannot change the integral. This gives selection rules for the nonvanishing of certain transitions or matrix elements.

For molecular, positive and negative center coincidence, which means the molecular dipole moment is equal to zero, and molecular is nonpolar. Molecules have the dipole moment, and the molecule is polar molecules. Dipole moment not only has size, and direction, being a vector, but also is a static physical quantity- the molecules of a static physical quantity in any symmetry under operation will not change. Whoever has the center of symmetry or with symmetry elements of the public node molecules has no dipole moment. In other cases, if only one C_n shaft, or only one plane of symmetry, or a C_n shaft is contained in one plane of symmetry inside, it may have dipole moment. For example, H_2O and NH_3 molecules have dipole moment, and are polar molecules. Although H_2O molecules have a C_2 shaft, it disjoints with two symmetry plane; NH_3 molecule has a C_3 shaft, but it is three symmetrical the intersecting line; CO_2 a center of symmetry I, so is a polar molecule; CCl_4 although no center of symmetry, but its four C_3 shaft and three C_2 shaft in the carbon atom place intersect one point, so permanent dipole moment is zero, molecular nonpolar. In a word, if the molecules belong to any of a group of points, it can't be polar molecules:

(1) contains inversion center group;

(2) any group D (including Dn, Dnh and Dnd)

(3) cube group (T, O), icosahedron group (I)

Messages consisting of sequences of binary digits (called words) transmitted over noisy channels are subject to errors. Although we cannot entirely eliminate the noisy channel from causing such errors, we can minimize the probability of them being misinterpreted at the receiving end by *encoding* the words.

Example 2.7.4　Suppose we want to send a message consisting of a sequence of two binary digits. The possible message words, each of *length* two , would be 00,10,01 and 11. If the probability of receiving the wrong word is $1-(0.99)^2 \doteq 0.02$. It is clear that the probability of receiving the wrong word increases with the length of the word and the length of the massage.

The basic idea is to adjoin to the words which carry the information (in this case, 00,10,01 and 11) *check digits* resulting in an *encoded word.* These digits are redundant in the sense that they do not carry any information but serve to detect and even correct errors. We think of the encoded words as consisting of two parts: the massage at the tail-end and the check digits at the beginning of the word as we read from left to right. Any collection of such words is called a *code.*

In this example above, let us adjoin a 0 or a 1 to each of the four words 00,01,10 and 11 to get the four words 000,101,110 and 011 with the property that each has an even number of 1's. Note that our original four words constituted the collection of all words of length two on the *alphabet* $\{0, 1\}$. On the other hand, the set of encoded words is a proper subset of the set of all words of length three. This is analogous to the situation in English where not all 26^4 "words" with four letters are words of the English language.

Suppose now that in the transmission of a message written in the code above, a single error occurs in one of the words. Then the received word will have an odd number of 1's and the receiver will detect that the word is not a word of the "language" and will ask for a re-transmission. This code is therefore called a *single-error-detecting* code. In fact the code will detect any odd number of errors but not an even number of errors.

Let us now compute the probability that a given transmitted word will be incorrectly interpreted at the receiving end. This will happen only when exactly two errors occur and the probability of this event is $\binom{3}{2}(0.01)^2(0.99) = 0.000297$, a considerable improvement over 0.02, the probability of error if the uncoded words are transmitted. As is always the case, we don't get something for nothing: for the sake of accuracy we have increased transmission time. A moment's thought will convince the reader that this will always be the case: as accuracy increases, efficiency decreases.

It is often the case that a message can only be sent once. For example, it would hardly be practical to stop a modern high-speed computer each time an error is detected. In such situations error-detecting codes are therefore of little value. Instead error-correcting codes are used. These are codes which detect errors and then correct them according to some prescribed procedure. Once again we illustrate with a simple example.

Suppose we want our language to consist of two words 0 and 1 where the probability of error is again 0.01. We encode the information words as 000 and 111 and stipulate the following *decoding scheme*: the received word is decoded according to majority rule, i.e., if the received word has more $x's$ than $y's$, it will be read as xxx. What is the probability that a decoding error will occur?

Suppose xxx is transmitted. It will be incorrectly decoded if there is an error in at least two of the coordinate positions. This will happen with probability $\binom{3}{2}(0.01)^2(0.99)+(0.01)^3 \doteq 0.000298$, an improvement over 0.01, the probability of error if the uncoded words are transmitted. Once again, we have paid for accuracy by tripling transmission time:

The diagram below illustrates schematically the situation we have just described:

Message $\longrightarrow$ **Encoder** $\longrightarrow$ **Received word** $\longrightarrow$ **Decoder** $\longrightarrow$ **Message**

The English language has considerable error-correcting capacity. For example, if we come across the word "depcribe", we would have no difficulty in recognizing that there is a typographical error in the third letter and would correct it to read "describe". The reason we are able to do this is that there are no words different from but "close enough" to "describe". If on the other hand the word "cup" mistakenly appears as "cap", then, unless the context tells us, we do not even detect the error. In this case "cup" is "close to" "cap".

We take our cue from this discussion and conjecture: more "farther apart" the words of a code are, more likely it will be that we shall be able to recognize and even correct errors. We shall prove a precise version of this conjecture after we have introduced the concept of Hamming distance between words.

Definition 2.7.5　A **code** is a subset of $\mathbb{Z}_2^n$. The elements of a code are called **words** and the number of coordinate positions (in this case n) is the **length** of a word.

Since $\mathbb{Z}_2^n$ is a ring, we make use of its algebraic structure so that, given two words α and β, we obtain a word $\alpha+\beta$.

Definition 2.7.6　If α is a word in $\mathbb{Z}_2^n$, we denote the ith coordinate of α by α_i.

(1) The **support** $S(\alpha)$ of α is the set $\{i \mid \alpha_i = 1\} \subseteq \{1, 2, \cdots, n\}$;

(2) The **weight** $w(\alpha)$ of α is $|S(\alpha)|$;

(3) If α and β are in $\mathbb{Z}_2^n$, the **Hamming distance** $h(\alpha, \beta)$ is the number of coordinate positions where α and β differ.

Let us assume that we can transmit messages only once. We must therefore devise some **complete decoding scheme**, that is, we must supply the receiver with an algorithm which will enable him/her to decode any received word, even if there are errors. This is in sharp contrast which obtains when we discussed the single-error-detecting code above where decoding does not take place at all if an error is detected.

Then, how do we devise such a decoding scheme? We shall be considering only channels where the probability of errors in any one symbol is independent of what happens to any other symbol. In such channels, it is less likely for $k+1$ errors to occur than for k to occur.

We therefore adopt the following scheme: if β is received, find the nearest word α in the adopted code and decode β as α. If there is more than one codeword closest to β, make an arbitrary choice from among the code words with minimum distance from β. Of course, if β is itself a code word, it will be decoded as β. This method of decoding is called the *nearest neighbor decoding.*

Suppose that we have chosen a code C and that the word α of this code is transmitted and β is received. It is clear that γ has ones in precisely those coordinate positions where errors have occurred. Using this terminology we may describe the nearest neighbor decoding as follows: find all error words and choose the code word with associated error word of the least weight.

Definition 2.7.7 (1) A code C is said to be ***d*-error detecting** if it recognizes all possible patterns of d or fewer errors; it is said to be ***d*-error detecting**, if using the nearest neighbor decoding scheme, it corrects all possible patterns of d or fewer errors.

(2) **The minimum Hamming distance** $\mu(C)$ *of a code* C is defined by $\mu(C) = \min\{h(\alpha,\beta) \mid \alpha,\beta \in C, \alpha \neq \beta\}$.

Example 2.7.8 If $C = 0000, 0110, 1001, 1111$, then $\mu(C) = 2$.

Example 2.7.9 Let $C = \{00000, 101100, 011010, 110001, 110110, 011101, 101011, 000111\}$. This code is a subgroup of $\mathbb{Z}_2^5$ and so it is closed under addition. Consequently, $\mu(C) = \min\{w(\alpha) \mid \alpha \in C, \alpha \neq 0\}$(Why). Therefore $\mu(C) = 3$ and so, by Definition2.7.7, C is a single-error-correcting code.

We will apply group theory to the construction of a certain class of codes. The notions of homomorphisms and subgroups will play prominent roles in the theoretical development while it will be seen that cosets can be effectively used in the practical process of decoding.

Definition 2.7.10 A **group code** is any subgroup of $(\mathbb{Z}_2^n, +)$.

Definition 2.7.11 Let $i_1, i_2, \cdots, i_k$ be a subset of $\{1, 2, \cdots, n\}$ where $i_u < i_v$ if $u < v$. Define the function $\Pi_s : \mathbb{Z}_2^n \to \mathbb{Z}_2^k$ as follows: if $\alpha = a_1 a_2 \cdots a_n$,

$$\Pi_s(\alpha) = a_{i_1} a_{i_2} \ldots a_{i_k}.$$

Example 2.7.12 Let $S = \{1,3\} \subset \{1,2,3,4\}$. Then $\Pi_s(1011) = 11$; $\Pi_s(1101) = 10$.

Example 2.7.13 Consider the group code $C = \{000, 001, 100, 101\}$. Permuting the first two coordinate positions we get $C' = \{0\dot{:}00, 0\dot{:}01, 0\dot{:}10, 0\dot{:}11\}$ where each element of $\mathbb{Z}_2^2$ appears on the right of the dotted line. In this case, if we take $S = \{1,3\}, \Pi_s(C) = \mathbb{Z}_2^2$.

Definition 2.7.14 An (n,k) group code C is a group code contained in $\mathbb{Z}_2^n$ such that

$$\Pi_T(C) = \mathbb{Z}_2^k \text{ where } T = \{n-k+1, n-k+2, \cdots, n\}.$$

In other words, every word of $\mathbb{Z}_2^k$ occurs in the last K coordinate positions of some word of C.

Example 2.7.15 {000000, 101100, 011010, 110001, 110110, 011101, 101011, 000111} is a $(6,3)$ group code. In this case, we start with the eight words of length three and adjoin check digits.

国际前沿研究动态

(1) 当 G 是一个有限群时, 由此可以定义一个 Hopf 代数 (Hopf algebra) 与 Hopf 群余代数 (Hopf group-coalgebra algebra), 研究见文献 [2]~[5]、[10].

(2) 当 G 是一个无限群时, 由此可以定义一个乘子 Hopf 代数 (multiplier Hopf algebra), 研究见文献 [1]、[6]~[9]、[11]~[16].

参 考 文 献

[1] El-hafez A T A, Delvaux L, Daele V. Group-cograded multiplier Hopf (*-)algebras, Alg. Represent Theory, 2007, 10: 77-95.

[2] Beattie M, Dăscălescu S, Grünenfelder L, et al. Finiteness conditions, co-Frobenius Hopf algebras, and quantum groups. J. Algebra, 1998, 200(1): 312-333.

[3] Caenepeel S, Lombaerde M D. A categorical approach to Turaev's Hopf group-coalgebras. Comm. Algebra, 2006, 34(7): 2631-2657.

[4] Caenepeel S, Janssen K, Wang S H. Group coring. Appl. Categor. Struct., 2008, 16(1-2): 65-96.

[5] Chari V, Pressley A. A Guide to Quantum Groups. New York: Cambridge University Press, 2010.

[6] Van Daele A. Multiplier Hopf algebras. Trans. Am. Math. Soc., 1994, 342(2): 917-932.

[7] Van Daele A. An algebraic framework for group duality. Adv. in Math., 1998, 140: 323-366.

[8] Van Daele A. The Fourier transform in quantum group theory. Proceedings of New Techniques in Hopf Algebras and Graded Ring Theory. 2007: 187-196.

[9] Van Daele A. Tools for working with multiplier Hopf algebras. Arab. J. Sci. Eng., 2008, 33(2C): 505-527.

[10] Van Daele A, Wang S H. New braided crossed categories and Drinfel'd quantum double for weak Hopf group coalgebras. Comm. Algebra, 2008, 36(6): 2341-2386.

[11] Van Daele A, Wang S H. Larson-Sweedler theorem and some properties of discrete type in (G-cograded) multiplier Hopf algebras. Comm. Algebra, 2006, 34(6): 2235-2249.

[12] Van Daele A, Wang S H. The Larson-Sweedler theorem for multiplier Hopf algebras. J. Algebra, 2006, 296: 75-95.

[13] Van Daele A, Wang S H. On the twisting and Drinfe'd double multiplier Hopf algebras. Comm. Algebra, 2006, 34(8): 2811-2842.

[14] Van Daele A, Wang S H. A class of multiplier Hopf algebras. Alg. Represent. Theory, 2007, 10: 441-461.

[15] Van Daele A, Wang S H. Pontryagin duality for bornological quantum hypergroups. Manuscripta Math., 2010, 131: 247-263.

[16] 王栓宏. Pontryagin 对偶与代数量子超群. 北京: 科学出版社, 2011.

习　题

1. True or false with reasons. Here, G is always a group.

(i) The function $e : \mathbb{N} \times \mathbb{N} \to \mathbb{N}$, defined by $e(m,n) = m^n$, is an associative operation.

(ii) Every group is Abelian.

(iii) The set of all positive real numbers is a group under multiplication.

(iv) The set of all positive real numbers is a group under addition.

(v) For all $a, b \in G$, where G is a group, it holds $aba^{-1}b^{-1} = 1$.

(vi) Every permutation of the vertices v_1, v_2, v_3 of an equilateral triangle π_3 is the restriction of a symmetry of π_3.

(vii) Every permutation of the vertices v_1, v_2, v_3, v_4 of a square π_4 is the restriction of a symmetry of π_4.

(viii) If $a, b \in G$, where G is a group, then $(ab)^n = a^n b^n$ for all $n \in \mathbb{N}$.

(ix) Every infinite group contains an element of infinite order.

(x) Complex conjugation permutes the roots of every polynomial having real coefficients.

(xi) If H is a subgroup of K and K is a subgroup of G, then H is a subgroup of G.

(xii) G is a subgroup of itself.

(xiii) The empty set $\varnothing$ is a subgroup of G.

2. If $a_1, a_2, \cdots, a_n$ are (not necessarily distinct) elements in a group G, prove that

$$(a_1 a_2 \cdots a_n)^{-1} = a_n^{-1} \cdots a_2^{-1} a_1^{-1}.$$

3. Let G be a group and let $a \in G$ have order dk, where $d, k > 1$. Prove that if there is $x \in G$ with $x^d = a$, then the order of x is $d^2 k$. Conclude that the order of x is larger than the order of a.

4. (i) How many elements of order 2 are there in S_5 and in S_6?

(ii) How many elements of order 2 are there in S_n?

5. Let G be a finite group in which every element has a square root, that is, for each $x \in G$, there exists $y \in G$ with $y^2 = x$. Prove that every element in G has a unique square root.

6. If G is a group with an even number of elements, prove that the number of elements in G of order 2 is odd. In particular, G must contain an element of order 2.

7. What is the largest order of an element in S_n, where $n = 1, 2, \cdots, 10$?

8. Give an example of two subgroups H and K of a group G whose union $H \cup K$ is not a subgroup of G.

9. Let G be a finite group with subgroups H and K. If $H \leqslant K$, prove that

$$[G : H] = [G : K][K : H].$$

10. If H and K are subgroups of a group G and if $|H|$ and $|K|$ are relatively prime, prove that $H \cap K = \{1\}$.

11. Prove that every infinite group contains infinitely many subgroups.

12. Let G be a finite group, and let S and T be (not necessarily distinct) nonempty subsets. Prove that either $G = ST$ or $|G| \geqslant |S| + |T|$.

13. If there is a bijection $f : X \to Y$ (that is , if X and Y have the same number of elements), prove that there is an isomorphism $\varphi : S_X \to S_Y$.

14. Let G be a group, X be a set, and $\varphi : G \to X$ be a bijection. Prove that there is an operation on X which makes X into a group such that $\varphi : G \to X$ is an isomorphism.

15. Prove that a group G is Abelian if and only if the function $f : G \to G$, given by $f(a) = a^{-1}$, is a homomorphism.

16. (i) Show that every group G with $|G| < 6$ is Abelian.

(ii) Find two nonisomorphic groups of order 6.

17. (i) Find a subgroup $H \leqslant S_4$ with $H \cong V$ but with $H \neq V$, where V is the Klein group.

(ii)Prove that the subgroup H in part(i) is not a normal subgroup.

18. If G is a group and $a, b \in G$, prove that ab and ba have the same order.

19. Let G be the additive group of all polynomials in x with coefficients in $\mathbb{Z}$, and let H be the multiplicative group of all positive rationals. Prove that $G \cong H$.

20. Show that if H is a subgroup with $bH = Hb = \{hb : h \in H\}$ for every $b \in G$, then H must be a normal subgroup.

21. Prove that the intersection of any family of normal subgroups of a group G is itself a normal subgroup of G.

22. Let G be a finite group written multiplicatively. Prove that if $|G|$ is odd, then every $x \in G$ has a square root.

23. Give an example of a group G, a subgroup $H \leqslant G$, and an element $g \in G$ with $[G : H] = 3$ and $g^3 \notin H$.

24. If G is a group, prove that $\mathbf{Aut}(G) = \{1\}$ if and only if $|G| \leqslant 2$, where $\mathbf{Aut}(G)$ represents all the isomorphisms of G to itself.

25. If C is a finite cyclic group of order n, prove that $|\mathbf{Aut}(C)| = \phi(n)$, where $\phi(n)$ is the Euler ϕ-function.

26. Let G be a finite group, p be prime, and let H be a normal subgroup of G. Prove that if both $|H|$ and $|G/H|$ are powers of p, then $|G|$ is a power of p.

27. Let A, B and C be groups, and let α, β and γ be homomorphisms with $\gamma \circ \alpha = \beta$.

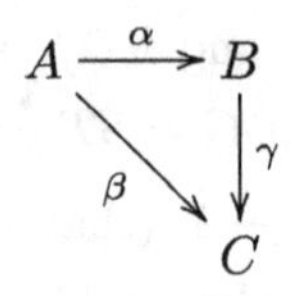

If α is surjective, prove that $\ker\gamma = \alpha(\ker\beta)$.

28. Let G be a finite group with $K \lhd G$. If $(|K|, [G:K]) = 1$, prove that K is the unique subgroup of G of order $|K|$.

29. Prove that if K is a subgroup of a group G, and if every left coset aK is equal to a right coset Kb, then $K \lhd G$.

第3章　环 (Rings)

本章主要介绍环 (ring) 的概念及其结构性质, 涉及子环 (subring)、理想 (ideal)、环同态和商环 (quotient ring). 它的深层次研究会涉及乘子环 (mulitiplier ring)、Hopf 代数的交叉积 (crossed product for Hopf algebras) 与余环 (coring) 理论的建立. 本章的研究与学习方法与第 2 章群理论的建立方法相同.

The study of rings originated from the theory of polynomial rings and the theory of algebraic integers. Furthermore, the appearance of hyper-complex numbers in the mid-19th century undercut the pre-eminence of fields in mathematical analysis. Commutative ring theory originated in algebraic number theory, algebraic geometry, and invariant theory. Central to the development of these subjects were the rings of integers in algebraic number fields and algebraic function fields, and the rings of polynomials in two or more variables. Noncommutative ring theory began with attempts to extend the complex numbers to various hyper-complex number systems. The genesis of the theories of commutative and noncommutative rings dates back to the early 19th century, while their maturity was achieved only in the third decade of the 20th century.

3.1　环　概　念

本节主要介绍环 (ring) 的基本概念与例子, 同时给出了环的一些基本性质.

Definition 3.1.1　A **ring** $(R,+,\cdot)$(环) is a set R with two binary operations$+$ and $\cdot$, which we call **addition** (加法) and **multiplication** (乘法), defined on R such that the following axioms are satisfied:

(1) $(R,+)$ is an Aabelian group with identity denoted by 0 and called the **zero** of the ring;

(2) Multiplication is associative;

(3) For all $a,b,c\in R$, $a\cdot(b+c)=a\cdot b+a\cdot c$ and $(b+c)\cdot a=b\cdot a+c\cdot a$. These are called the **left** and **right distributivity laws**(分配律) respectively.

We have not insisted that a ring posses a multiplication identity. When it does, we shall denote that identity by 1 and postulate $1 \neq 0$. We shall refer to the ring as a **ring with unity**. To distinguish the multiplicative identity from the additive identity, we call the former the **unit element** or the **unity** (单位元) of the ring.

When the multiplication is commutative, we call the ring a **commutative ring** (交换环).

It is customary to denote multiplication in a ring by juxtaposition, using ab in place of $a \cdot b$. Also, we can refer to a **ring** R in place of a **ring** $(R, +, \cdot)$, provided that no confusion will result.

Example 3.1.2 (1) The integers under ordinary addition and multiplication form a commutative ring with unity which we denote by $(\mathbb{Z}, +, \cdot)$.

(2) The set of even integers under ordinary addition and multiplication is a commutative ring (without unity).

Example 3.1.3 Let R be any ring and let $M_n(R)$ be the collection of all $n \times n$ matrices having elements of R as entries. The operations of addition and multiplication in R allow us to add and multiply in the usual fashion. We can quickly check that $(M_n(R), +)$ is an Abelian group. The associativity of matrix multiplication and the two distributive laws in $M_n(R)$ are straightforward calculations which indicate that they follow from the same properties in R. Thus we know that $M_n(R)$ is a ring. In particular, we have the rings $M_n(\mathbb{Z})$, $M_n(\mathbb{Q})$, $M_n(\mathbb{R})$ and $M_n(\mathbb{C})$. Note that multiplication is not a commutative operation in any of these rings for $n \geqslant 2$.

Example 3.1.4 (1) Recall that in group theory, $n\mathbb{Z}$ is the cyclic subgroup of $\mathbb{Z}$ under addition consisting of all integer multiples of the integer n. Since $(nr)(ns) = n(nrs)$, we see that $n\mathbb{Z}$ is closed under multiplication. The associative and distributive laws which hold in $\mathbb{Z}$ then assure us that $(n\mathbb{Z}, +, \cdot)$ is a ring. From now on we will consider $n\mathbb{Z}$ to be this ring.

(2) Consider the cyclic group $(\mathbb{Z}_n, +)$. If we define for $a, b \in \mathbb{Z}_n$, the product ab as the remainder of the usual product of integers when divided by n, it can be shown that $(\mathbb{Z}_n, +, \cdot)$ is a ring. For example, in $\mathbb{Z}_{10}$ we have $\overline{3} \cdot \overline{7} = \overline{1}$. This operation on $\mathbb{Z}_n$ is **multiplication modulo** n. From now on, $\mathbb{Z}_n$ will always be the ring $(\mathbb{Z}_n, +, \cdot)$.

Example 3.1.5 Let G be any Abelian group and $E = End(G)$ be the collection of all endomorphisms of G. For any $\sigma, \tau \in E, x \in G$, if we define

$$(\sigma + \tau)(x) = \sigma(x) + \tau(x), \tag{3.1}$$

$$(\sigma \cdot \tau)(x) = \sigma(\tau(x)), \tag{3.2}$$

then $(E, +, \cdot)$ forms a ring (called the **endomorphism ring of** G(G 的自同态环)).

证明 设 $\sigma, \tau \in E$, 则由 (3.2) 式可知 $\sigma \cdot \tau$ 是 G 的自同态. 显然 $\sigma + \tau$ 是 G 到自身的映射, 且对任意的 $x, y \in G$,

$$\begin{aligned}(\sigma + \tau)(x + y) &= \sigma(x + y) + \tau(x + y)\\ &= \sigma(x) + \sigma(y) + \tau(x) + \tau(y)\\ &= \sigma(x) + \tau(x) + \sigma(y) + \tau(y)\\ &= (\sigma + \tau)(x) + (\sigma + \tau)(y),\end{aligned}$$

从而 $\sigma + \tau \in E$. 因此, $+, \cdot$ 是 E 上的两个运算.

容易证明 $+$ 满足结合律与交换律, 且零同态 0(即将 G 中每一个元素都映为 G 的零元) 为 E 的零元. 对任意的 $\sigma \in E$, 令

$$(-\sigma)(x) = -\sigma(x), \forall x \in G,$$

则 $-\sigma \in E$, 显然 $-\sigma$ 是 σ 在 E 中的负元. 从而 $(E, +)$ 为一个加群.

易证 $\cdot$ 满足结合律. 最后, 对任意的 $\sigma, \tau, \lambda \in E, x \in G$, 有

$$\begin{aligned}[(\sigma + \tau) \cdot \lambda](x) &= (\sigma + \tau)[\lambda(x)]\\ &= \sigma[\lambda(x)] + \tau[\lambda(x)]\\ &= (\sigma \cdot \lambda)(x) + (\tau \cdot \lambda)(x)\\ &= (\sigma \cdot \lambda + \tau \cdot \lambda)(x),\end{aligned}$$

从而 $\cdot$ 关于 $+$ 满足右分配律. 同理可证, $\cdot$ 关于 $+$ 满足左分配律. 因此 $(E, +, \cdot)$ 是一个环. □

Theorem 3.1.6 If R is a ring with additive identity 0, then for any $a, b \in R$, we have

(1) $0a = a0 = 0$.

(2) $a(-b) = (-a)b = -(ab)$.

(3) $(-a)(-b) = ab$.

证明 (1) 对于 $a \in R$, 有

$$a0 + a0 = a(0 + 0) = a0 = 0 + a0,$$

由加法右消去律可得 $a0 = 0$. 同理 $0a = 0$.

(2) 对于 $a, b \in R$, 利用左分配律和 (1), 有

$$a(-b) + ab = a(-b + b) = a0 = 0,$$

即 $a(-b) = -(ab)$. 同理 $(-a)b = -(ab)$.

(3) 对于 $a, b \in R$, 利用 (2), 有

$$(-a)(-b) = -[a(-b)] = -[-(ab)] = ab. \qquad \square$$

We shall write $a-b$ instead of $a+(-b)$ and call "-" **subtraction**(减法). It is easy to show that multiplication is distributive over subtraction. Note that subtraction is not an associate operation.

Theorem 3.1.7 Let R be a ring with left multiplication identity e. If e is the sole left multiplication identity in R, then e is the unit element of R.

证明 只需证明 e 是 R 的右乘法单位元. 用反证法, 若 e 不是 R 的右乘法单位元, 则必存在 $x \in R$, 使得 $xe \neq x$, 则 $e + xe - x \neq e$. 但对于任意的 $y \in R$, 有

$$(e + xe - x)y = y + xy - xy = y,$$

即 $e + xe - x$ 是 R 的另一个左乘法单位元, 与题设相矛盾. 故 e 为 R 的单位元. □

Definition 3.1.8 (1) Let R be a ring with unity. An element $r \in R$ is a **unit** (单位) if r has a multiplicative inverse in R.

(2) An element a of a ring R is a **left zero divisor** (左零因子) if there is a nonzero element $b \in R$ such that $ab = 0$, and is a **right zero divisor** (右零因子) if there is a nonzero element $c \in R$ such that $ca = 0$.

If $ab = 0$ in a ring R, then a is a left zero divisor and b is a right zero divisor. Thus the left and right zero divisors come in pairs. We call left zero divisor and right zero divisor the **zero divisor** (零因子)for short.

In a ring with unity, it can be showed that the set of units forms a group under the multiplicative operation of the ring. This group is called the **group of units of the ring**.

Be careful not to confuse our use of the words **unit**(单位) and **unity**(单位元).

Example 3.1.9 (1) $2 \in \mathbb{Z}$ is not a unit since 2 has no multiplicative inverse. The only units in $\mathbb{Z}$ are 1 and -1.

(2) The group of units of the ring $\mathbb{R}$ of real numbers consists of $\mathbb{R}^*$, the set of nonzero real numbers.

Example 3.1.10 Consider the ring R of 2×2 upper triangular matrices with real entries. The matrix $\begin{pmatrix} a & b \\ 0 & c \end{pmatrix}$ is a unit of this ring if and only if $ac \neq 0$.

Example 3.1.11 Let us find the units in $\mathbb{Z}_{14}$. Of course, 1 and $-1 = \overline{13}$ are units. Since $\overline{3} \cdot \overline{5} = 1$, we see that $\overline{3}$ and $\overline{5}$ are units; therefore $-\overline{3} = \overline{11}$ and $-\overline{5} = \overline{9}$ are also units. None of the remaining elements of $\mathbb{Z}_{14}$ can be units, since no multiple of $\overline{2}$, $\overline{4}$, $\overline{6}$, $\overline{7}$, $\overline{8}$ or $\overline{10}$ can be one more than a multiple of 14; they all have a common factor, either $\overline{2}$ or $\overline{7}$, with 14. That means they are all zero divisors. Furthermore, it can be showed that the units in $\mathbb{Z}_n$ are precisely those $m \in \mathbb{Z}_n$ such that $\gcd(m, n) = 1$.

Example 3.1.12 Let $A \in M_n(\mathbb{R})$. Then it is easy to check that A is a unit of ring $M_n(\mathbb{R})$ if and only if $|A| \neq 0$. Thus the group of units of the ring $M_n(\mathbb{R})$ is $GL_n(\mathbb{R})$, the group of all invertible matrices in $M_n(\mathbb{R})$.

Proposition 3.1.13 Let R be any ring, then R contains no left(right) zero divisions if and only if the cancellation laws hold in R.

证明 设 R 无左零因子. 令 $a \neq 0$, 若 $ab = ac$, 则

$$ab - ac = a(b - c) = 0,$$

于是 $b - c = 0$, 即 $b = c$. 因此左消去律成立.

由于左右零因子总是成对出现, 因此若 R 无左零因子, 则 R 中必无右零因子. 从而右消去律同理可证.

反之, 设 R 中乘法消去律成立, 令 a 为 R 中非零元素, 若存在 $b \in F$ 使得 $ab = 0 = a0$, 则由消去律可得 $b = 0$, 即 a 非左零因子. 同理可证 R 中无右零因子. □

Definition 3.1.14 Let R be any ring with unity 1. If every nonzero element of R is a unit, then R is a **division ring** (or **skew field**) (除环). A **field** (域) is a commutative division ring.

Definition 3.1.15 Let R be a commutative ring with unity 1. If R contains no zero divisions, then R is a **domain** (or an **integral domain**) (整环).

Example 3.1.16 From Example 3.1.9 we can see that $\mathbb{Z}$ is not a field but an integral domain; $\mathbb{R}$ is a field.

Example 3.1.17 From Example 3.1.11 we can see that $\mathbb{Z}_p$ for any prime p is an integral domain, but $\mathbb{Z}_n$ is not an integral domain if n is not prime.

Theorem 3.1.18 Every field F is an integral domain.

证明 只需证明 F 中没有零因子即可. 设非零元 $a \in F$, 若存在 $b \in F$ 使得 $ab = 0$, 则

$$a^{-1}ab = a^{-1}0 = 0.$$

另一方面

$$a^{-1}ab = 1b = b,$$

即 $b = 0$. 因此 F 中无零因子. 从而 F 必为整环. □

3.2 子环 (Subrings) 与环同态

本节介绍子环与环同态的概念与基本性质.

Definition 3.2.1 A subset S is a **subring** (子环) of a ring R if it is a ring in its own right under the operations inherited from R.

Example 3.2.2 R is a ring, then 0 and R are subrings of R. We call them the **trivial subrings** (平凡子环)of R.

Example 3.2.3 The set of even integers is a subring of the ring of integers. But the former has no unity while the latter has one.

Example 3.2.4 Let R be any ring. Then the ring $M_n(R)$ contains the subrings of upper triangular and strictly upper triangular matrices.

Example 3.2.5 $R' = \left\{ \begin{pmatrix} a & 0 \\ 0 & 0 \end{pmatrix} \mid a \in \mathbb{R} \right\}$ and $R'' = \left\{ \begin{pmatrix} 0 & 0 \\ 0 & b \end{pmatrix} \mid b \in \mathbb{R} \right\}$ are two subrings of $M_2(\mathbb{R})$.

The unity in R' is $\begin{pmatrix} 1 & 0 \\ 0 & 0 \end{pmatrix}$ and the unity in R'' is $\begin{pmatrix} 0 & 0 \\ 0 & 1 \end{pmatrix}$; note that they are different with the unity of $M_2(\mathbb{R})$.

Theorem 3.2.6 A subset S is a subring of ring R if and only if

(1) $S \neq \varnothing$;

(2) $s, t \in S$ implies $s - t \in S$;

(3) $s, t \in S$ implies $st \in S$.

证明 若 S 为环 R 的子环, 则显然 S 为 R 的加法子群. 于是条件 (1) 和 (2) 成立. 又由 S 对乘法封闭, 可知条件 (3) 成立.

另一方面, 若条件 (1) 和 (2) 成立, 则 S 为 R 的加法子群. 并且条件 (3) 成立说明 S 对乘法封闭, 此时由 R 中乘法结合律与左右分配律成立可知, 在 S 中乘法结合律与左右分配律亦成立. 于是 S 为环 R 的子环. □

Definition 3.2.7 For rings R and S, a map $f : R \to S$ is a **homomorphism** if the following two conditions are satisfied for all $a, b \in R$:

(1) $f(a + b) = f(a) + f(b)$;

(2) $f(ab) = f(a)f(b)$.

A one-to-one ring homomorphism is a **ring monomorphism** (环单态); an onto ring homomorphism is a **ring epimorphism** (环满态); an one-to-one onto ring homomorphism is a **ring isomorphism** (环同构). As groups, we write $R \cong S$ if there is an isomorphism between R and S and we say that the two rings are **isomorphic**.

In this definition, Condition (1) is the statement that f is a homomorphism mapping the Abelian group $(R, +)$ to $(S, +)$. Condition (2) requires that f relates the multiplicative structure of the rings R and S in the same way. Since f is also a group homomorphism, all the results concerning group homomorphisms are valid for the additive structure of the rings. For example, $f : R \to S$ is a ring monomorphism if and only if its **kernel**($\ker f = \{a \in R \mid f(a) = 0\}$) is just the subset $\{0\}$ of R; f is a ring epimorphism if and only if its **image**($\operatorname{Im} f = \{f(a) \in S \mid a \in R\}$) is S.

Example 3.2.8 Complex conjugation, defined by $z = a + ib \mapsto \overline{z} = a - ib$, is a ring homomorphism $\mathbb{C} \to \mathbb{C}$.

Example 3.2.9 The map $\phi : \mathbb{Z} \to \mathbb{Z}_n$ where $\phi(a)$ is the remainder of a modulo n is a ring homomorphism for each positive integer n. We know $\phi(a+b) = \phi(a)+\phi(b)$ by group theory. To show the multiplicative property, write $a = q_1n+r_1$ and $b = q_2n+r_2$ according to the division algorithm. Then $ab = n(q_1q_2n + r_1q_2 + q_1r_2) + r_1r_2$. Thus $\phi(ab)$ is the remainder of r_1r_2 when divided by n. Since $\phi(a) = r_1$, $\phi(b) = r_2$, Example

3.1.4 (2) indicates that $\phi(a)\phi(b)$ is also the same remainder, so $\phi(ab) = \phi(a)\phi(b)$. Moreover, ϕ is an epimorphism.

Example 3.2.10 As Abelian groups, $(\mathbb{Z},+)$ and $(2\mathbb{Z},+)$ are isomorphic under the map $f : \mathbb{Z} \to \mathbb{Z}$, with $f(x) = 2x$ for $x \in \mathbb{Z}$. Note that f here is not a ring isomorphism for $f(xy) = 2xy$ while $f(x)f(y) = 4xy$.

Theorem 3.2.11 Let $f : R \to S$ be a ring homomorphism. Then

(1) $f(0_R) = 0_S$ where 0_R and 0_S are the zeroes of R and S, respectively.

(2) If R admits unity 1_R, then $\mathrm{Im}\, f$ has unity $f(1_R)$. Thus if f is an epimorphism, the unity of S is $f(1_R)$.

(3) If R' is a subring of R, then $f(R')$ is a subring of S.

(4) If S' is a subring of S, then $f^{-1}(S') = \{a \in R \mid f(a) \in S'\}$ is a subring of S.

证明 (1) 直接可得.

(2) 对于任意的 $f(a) \in \mathrm{Im}\, f$, 有

$$f(a) = f(a1_R) = f(a)f(1_R) = f(1_Ra) = f(1_R)f(a),$$

从而 $f(1_R)$ 为 $\mathrm{Im}\, f$ 的单位元.

(3) 首先易知 $f(R')$ 非空. 又对任意的 $f(a), f(b) \in f(R')$, $f(a) - f(b) = f(a-b)$. 从而由 $a - b \in R'$ 可知 $f(a) - f(b) \in f(R')$. 同理可得 $f(a)f(b) \in f(R')$. 于是由定理 3.2.6 可知, $f(R')$ 为 S 的一个子环.

(4) 首先易知 $f^{-1}(S')$ 非空. 又对任意的 $a, b \in f^{-1}(S')$, 有 $f(a-b) = f(a) - f(b)$. 由 S' 为子环知 $f(a) - f(b) \in S'$, 从而 $a - b \in f^{-1}(S')$. 同理可得 $ab \in f^{-1}(S')$. 于是由定理 3.2.6 可知, $f^{-1}(S')$ 为 R 的一个子环. □

Theorem 3.2.12 Let $f : R \to S$ and $g : S \to T$ be ring homomorphisms. Then $g \circ f : R \to T$ is also a ring homomorphism.

证明 显然 $g \circ f$ 是 R 到 T 的映射. 对任意的 $x, y \in R$, 有

$$\begin{aligned}(g \circ f)(x + y) &= g(f(x + y)) = g(f(x) + f(y)) \\ &= g(f(x)) + g(f(y)) \\ &= (g \circ f)(x) + (g \circ f)(y),\end{aligned}$$

$$\begin{aligned}(g \circ f)(xy) &= g(f(xy)) = g(f(x)f(y)) \\ &= g(f(x))g(f(y)) \\ &= (g \circ f)(x)(g \circ f)(y).\end{aligned}$$

因此 $g \circ f$ 是环同态. □

Corollary 3.2.13 The ring isomorphism is an equivalence relation on the collection of rings.

3.3 理想 (Ideals) 与商环 (Quotient Rings)

本节主要介绍理想与商环.

Definition 3.3.1 Let R be a ring. A subring I of R is called a **left ideal** (左理想)(of R) if for all $r \in R$ and for all $x \in I$, $rx \in I$, and called a **right ideal** (右理想) (of R) if for all $r \in R$ and for all $x \in I$, $xr \in I$. If I is both the left and right ideal of a ring R, then I is called an **ideal** (理想) of R. One writes $I \lhd R$ in this case. If $I \neq R$, then I is a **proper ideal** (真理想) of R.

Since an ideal is, in particular, a subring, to show that I is an ideal of R we use the criteria established in Theorem 3.2.6, except that, instead of just showing closure under multiplication, we show that the product of an element of I with any element of R is in I. Therefore to prove I is an ideal of R we show

(1) $I \neq \varnothing$;

(2) if $x, y \in I$, then $x - y \in I$;

(3) if $r \in R$ and $x \in I$, then $rx \in I$ and $xr \in I$.

An ideal in ring theory is analogous to a normal subgroup in group theory. Both are the type of substructure needed to form a quotient structure. Think of an ideal as a "black hole" which draws into it any product of an element of the ideal with an element of the ring.

Example 3.3.2 We can see that $n\mathbb{Z}$ is an ideal of the ring $\mathbb{Z}$ since we know it is a subring, and $s(nm) = (nm)s = n(ms) \in n\mathbb{Z}$ for all $s \in \mathbb{Z}$.

Example 3.3.3 In the Example 3.2.3, the set of even integer is an ideal of $\mathbb{Z}$.

Example 3.3.4 R is a ring with unity, $a_1, a_2, \ldots, a_n \in R$. The set

$$I = \{r_1a_1s_1 + r_2a_2s_2 + \cdots + r_na_ns_n \mid r_i, s_i \in R, i \in \mathbb{Z}\}$$

is an ideal in R, and it is generated by $a_1, a_2, \ldots, a_n$. One writes $I = \langle a_1, a_2, \ldots, a_n\rangle$

in this case. In particular, if $n = 1$, then $I = \langle a \rangle$ is the ideal generated by a, and it is called the **principal ideal** (主理想).

Notice that if R admits unity 1, then R and $\{0\}$ are always principal ideals: $R = \langle 1 \rangle$ and $\{0\} = \langle 0 \rangle$. In $\mathbb{Z}$, the even integrals form the principal ideal $\langle 2 \rangle$.

Example 3.3.5 Let

$$I = \left\{ \begin{pmatrix} 0 & 0 & 0 \\ 0 & 0 & a \\ 0 & 0 & 0 \end{pmatrix} \mid a \in \mathbb{R} \right\}, \quad N = \left\{ \begin{pmatrix} 0 & 0 & x \\ 0 & 0 & y \\ 0 & 0 & 0 \end{pmatrix} \mid x, y \in \mathbb{R} \right\},$$

and

$$R = \left\{ \begin{pmatrix} a_1 & a_2 & a_3 \\ 0 & a_4 & a_5 \\ 0 & 0 & a_6 \end{pmatrix} \mid a_i \in \mathbb{R} \right\}.$$

Then it is easy to check that $I \lhd N$, $N \lhd R$, but I is not the ideal of R.

Theorem 3.3.6 Let R be a ring, and I an ideal of R. Then I is a normal subgroup of $(R, +)$ since $(R, +)$ is Abelian. Form the quotient group $(R/I, +)$. On R/I define an operation denoted by $\cdot$ (or juxtaposition) as follows:

$$\overline{a} \cdot \overline{b} = (a + I) \cdot (b + I) = ab + I = \overline{ab}, \quad \forall a + I, b + I \in R/I.$$

Then $(R/I, +, \cdot)$ is a ring called the **quotient ring of R mod I**. Moreover, the map $f : R \to R/I$ defined by $f(r) = r + I$ is an (ring) epimorphism (called the **natural homomorphism** or the **canonical homomorphism**) (自然同态) whose kernel is I.

证明 首先证明这个乘法的定义是合理的, 即乘积结果与陪集中代表元的选取无关. 设 $a + I = c + I, b + I = d + I$, 则 $a - c, b - d \in I$. 由 I 为理想, 有

$$(a - c)b \in I, \quad c(b - d) \in I,$$

从而 $(a-c)b+c(b-d) = ab-cd \in I$. 故 $ab+I = cd+I$, 即 $(a+I)\cdot(b+I) = (c+I)\cdot(d+I)$.

下证 $(R/I, +, \cdot)$ 为环. 首先易知 R/I 关于乘法满足结合律, 只需证明左右分配

律成立. 事实上, 对任意的 $a, b, c \in R$, 有

$$
\begin{aligned}
(a+I)(b+I+c+I) &= (a+I)(b+c+I) \\
&= a(b+c)+I \\
&= ab+ac+I \\
&= ab+I+ac+I \\
&= (a+I)(b+I)+(a+I)(c+I),
\end{aligned}
$$

即满足左分配律. 同理可证亦满足右分配律. 因此 $(R/I, +, \cdot)$ 是一个环.

现在考虑 $f: R \to R/I$. 对任意的 $a, b \in R$, $f(ab) = ab + I = (a+I)(b+I) = f(a)f(b)$, 即 f 保持乘法运算; 又由群论中结果可知 f 为加群同态, 因此 f 为环同态. 它的核显然为 I. □

Example 3.3.7 Example 3.3.2 shows that $n\mathbb{Z}$ is an ideal of $\mathbb{Z}$, so we can form the quotient ring $\mathbb{Z}/n\mathbb{Z}$.

Theorem 3.3.8 If $f: R \to S$ is a ring homomorphism, then

(1) if f is an epimorphism, and I is an ideal of R, then $f(I) \lhd S$;

(2) if J is an ideal of S, then $f^{-1}(J) \lhd R$. Furthermore, $\ker f = f^{-1}(\{0\})$ is an ideal since $\{0\}$ is an ideal of S.

证明 (1) 事实上对任意的 $f(a) \in f(I), s \in S$, 由 f 为满同态, 知存在 $r \in R$, 使得 $f(r) = s$. 从而

$$f(a)s = f(a)f(r) = f(ar).$$

又由 I 为理想可知, $ar \in I$, 即 $f(a)s \in f(I)$. 同理可得 $sf(a) \in f(I)$.

(2) 类似 (1) 可证. □

Definition 3.3.9 A **maximal ideal** (极大理想) of a ring R is an ideal I different from R such that there is no proper ideal J of R properly containing I.

Thus, if I is a maximal ideal, and if J is a proper ideal with $I \subseteq J$, then $I = J$.

Example 3.3.10 $p\mathbb{Z}$ is a maximal ideal of $\mathbb{Z}$ for any prime p.

Theorem 3.3.11 Let R be a commutative ring with unity. Then I is a maximal ideal of R if and only if R/I is a field.

证明 设 I 为极大理想, 易知 R/I 是有单位元 $1+I$ 的交换环, 故只需证明任意非零元 $a+I \in R/I$ 可逆即可.

令 $N = \{i + ax \mid i \in I, x \in R\}$, 易证 N 为 R 的一个理想, 且 $I \subsetneqq N$. 由 I 为极大理想, 可知 $N = R$. 于是 $1 \in N$, 即存在 $i \in I, x \in R$ 使得 $1 = i + ax$. 因而有 $1 + I = i + ax + I = ax + I = (a + I)(x + I)$, 即 $a + I$ 可逆.

反之, 若 R/I 为域, 设 N 为 R 的一个理想, 且 $I \subsetneqq N$, 则必存在 $a \in N$, 但 $a \notin I$. 下证 $N = R$. 由 R/I 为域可知, $a + I$ 可逆. 从而存在 $x \in R$ 使得 $1 + I = (a + I)(x + I)$. 故 $1 - ax \in I \subset N$. 由 $ax \in N$ 可得 $1 \in N$. 进而 $N = R$. □

Definition 3.3.12　An ideal $I \neq R$ in a commutative ring R is a **prime ideal** (素理想)if $ab \in I$ implies that either $a \in I$ or $b \in I$ for any $a, b \in R$.

Recall that a nonzero commutative ring R is a domain if and only if $ab = 0$ in R implies $a = 0$ or $b = 0$. Thus, the ideal $\{0\}$ is a prime ideal of R if and only if R is a domain.

Example 3.3.13　$p\mathbb{Z}$ is a prime ideal of $\mathbb{Z}$ for any prime p.

Theorem 3.3.14　Let R be a commutative ring with unity. Then I is a prime ideal of R if and only if R/I is an integral domain.

证明　设 I 为素理想, 若在 R/I 中有 $(a + I)(b + I) = \overline{0}$, 即 $ab \in I$, 从而 $a \in I$ 或 $b \in I$, 即 $a + I = \overline{0}$ 或 $b + I = \overline{0}$. 从而 R/I 中无零因子. 易知 R/I 为有单位元 $1 + I$ 的交换环, 从而 R/I 为整环.

反之, 设 R/I 为整环. 对任意的 $a, b \in R$, 若 $ab \in I$, 即 $(a + I)(b + I) = \overline{0}$, 由 R/I 中无零因子, 知 $a + I = 0$ 或 $b + I = 0$, 即 $a \in I$ 或 $b \in I$. 于是 I 为素理想. □

Corollary 3.3.15　Every maximal ideal in a commutative ring R with unity is a prime ideal.

3.4　环同态基本定理 (The Fundamental Theory of Ring Homomorphisms)

本节主要证明三个环同态基本定理, 证明类似群的情况.

Theorem 3.4.1 (The First Fundamental Homomorphism Theorem)　Let $f : R \to S$ be a ring homomorphism with kernel N. Then $f(R)$ is a ring, and the map $\overline{f} : R/N \to f(R)$ given by $\overline{f}(x + N) = f(x)$ is an isomorphism.

证明　易知 $f(R)$ 为环, 且 N 为 R 的理想. 因此 $\overline{f}$ 的定义成立. 由群同态基本定理可知 $\overline{f}$ 为双射, 且保持加法运算. 因此只需证明 $\overline{f}$ 保持乘法即可.

事实上, $\overline{f}((x+N)(y+N)) = f(x)f(y) = f(xy) = \overline{f}(x+N)\overline{f}(y+N)$. 因此 $\overline{f}$ 为环同构. □

Corollary 3.4.2 Let $f: R \to S$ be a ring epimorphism with kernel N. Then R/N and S are isomorphic.

Example 3.4.3 From Example 3.3.7 we know that there is a quotient ring $\mathbb{Z}/n\mathbb{Z}$. Example 3.2.9 shows that $\phi: \mathbb{Z} \to \mathbb{Z}_n$ where $\phi(m)$ is the remainder of m modulo n is a homomorphism, and we see that $\ker \phi = n\mathbb{Z}$. Then the above corollary shows that the map $\overline{\phi}: \mathbb{Z}/n\mathbb{Z} \to \mathbb{Z}_n$ where $\overline{\phi}(m+n\mathbb{Z})$ is the remainder of m modulo n is well defined and is an isomorphism.

In summary, every ring homomorphism with domain R gives rise to a quotient ring R/N, and every quotient ring R/N gives rise to a homomorphism mapping R into R/N.

Lemma 3.4.4 If R is a ring, S is a subring and J is an ideal of R, then $S+J=\{s+j \mid s \in S, j \in J\}$ is a subring of R.

证明 只需证明 $S+J$ 对乘法封闭即可. 事实上, 由 J 为理想可得

$$(s+j)(s'+j') = ss' + sj' + js' + jj' \in S+J.$$ □

Theorem 3.4.5 (The Second Fundamental Homomorphism Theorem) Let R, S and J be as in the above lemma. Then

$$(S+J)/J \cong S/(S \cap J).$$

证明 由群同态基本定理, 我们可构造态射 $\theta: S \to (S+J)/J$ 使得 $\theta(s) = s+J$ 且核为 $S \cap J$. 易知 θ 为加群同态. 只需证明 θ 保持乘法即可. 而事实上 $\theta(ss') = ss'+J = (s+J)(s'+J) = \theta(s)\theta(s')$. 于是由环同态第一基本定理可知定理成立. □

Theorem 3.4.6 (The Third Fundamental Homomorphism Theorem) Let I and J be ideals of a ring R with $I \subseteq J$, then

$$(R/I)/(J/I) \cong R/J.$$

证明 证法同上. 具体证明留作练习. □

3.5 几类重要环

本节介绍几类重要环及其性质.

Definition 3.5.1 Let R be a ring. A **polynomial** $f(x)$ **with coefficients** in R (多项式)is an infinite formal sum

$$\sum_{i=0}^{\infty} a_i x^i = a_0 + a_1 x + \cdots + a_n x^n + \cdots,$$

where $a_i \in R$ and $a_i = 0$ for all but a finite number of values of i. The a_i are **coefficients of** $f(x)$ (系数). If for some $i \geqslant 0$ it is true that $a_i \neq 0$, the largest such value of i is the **degree of** $f(x)$ (次数). If all $a_i = 0$, then the degree of $f(x)$ is undefined.

We denote $f(x)$ by $a_0 + a_1 x + \cdots + a_n x^n$ to simplify our working if $f(x) = a_0 + a_1 x + \cdots + a_n x^n + \cdots$ has $a_i = 0$ for $i > n$.

Addition and multiplication of polynomials with coefficients in a ring R are defined as follows. If

$$f(x) = a_0 + a_1 x + \cdots + a_n x^n + \cdots$$

and

$$g(x) = b_0 + b_1 x + \cdots + b_n x^n + \cdots,$$

then for polynomial addition, we have

$$f(x) + g(x) = c_0 + c_1 x + \cdots + c_n x^n + \cdots,$$

where $c_i = a_i + b_i$. And for polynomial multiplication, we have

$$f(x)g(x) = d_0 + d_1 x + \cdots + d_n x^n + \cdots,$$

where $d_n = \sum_{j=0}^{n} a_j b_{n-j}$.

Then the set $R[x]$ of all polynomials in an indeterminate x with coefficients in a ring R is a ring under the above addition and multiplication, and we call it the **polynomial ring** (多项式环). R is a subring of $R[x]$.

Note that if R is commutative, then so is $R[x]$, and if R has unity 1, then 1 is also unity for $R[x]$.

Example 3.5.2 $\mathbb{Z}[x]$ is the ring of polynomials in the indeterminate x with integral coefficients; $\mathbb{Q}[x]$ is the ring of polynomials in the indeterminate x with rational coefficients, and so on.

Example 3.5.3 In $\mathbb{Z}_2[x]$, we have

$$(x+1)^2 = (x+1)(x+1) = x^2 + (1+1)x + 1 = x^2 + 1$$

and we obtain

$$(x+1) + (x+1) = (1+1)x + (1+1) = 0x + 0 = 0.$$

Theorem 3.5.4 R is an integral domain if and only if $R[x]$ is an integral domain.

证明 只需证明当 R 无零因子时, $R[x]$ 也无零因子. 任取两个非零多项式 $f(x)$ 与 $g(x)$, 且令

$$f(x) = a_0 + a_1x + \cdots + a_mx^m, \quad a_m \neq 0,$$

$$g(x) = b_0 + b_1x + \cdots + b_nx^n, \quad b_n \neq 0,$$

则由于 R 无零因子, $a_mb_n \neq 0$, 故

$$f(x)g(x) = a_0b_0 + \cdots + a_mb_nx^{m+n} \neq 0,$$

即 $R[x]$ 无零因子. □

Definition 3.5.5 Let R be a commutative ring with identity, and let $a, b \in R$. If $a \neq 0$, we say that a **divides** b and write $a \mid b$ if there exists $c \in R$ such that $b = ac$. If c is a unit of R, then a is said to be an **associate** (相伴元) of b.

If $a \mid b$ and a is neither an associate of b nor a unit in R, then a is called a **proper divisor**(真因子) of b.

Theorem 3.5.6 Let F be a field, and $f(x), g(x) \in F[x]$, $f(x) \neq 0$. Then there exist unique polynomials $q(x), r(x) \in F[x]$ such that

(1) $g(x) = q(x)f(x) + r(x)$;

(2) $\deg(r) < \deg(f)$, where $\deg(r)$ $(\deg(f))$ means the degree of $r(x)$ $(f(x))$.

证明 若 $f \mid g$, 则显然命题成立. 设 $f \nmid g$, 则考虑形如 $g - qf$ 的多项式中次数最低的那个, 设为 $r(x)$, 其中 $q(x)$ 取遍 $F[x]$. 从而有 $g(x) = q(x)f(x) + r(x)$. 现在只需证明 $\deg(r) < \deg(f)$.

设 $f(x)=s_0+s_1x+\cdots+s_nx^n$, $g(x)=t_0+t_1x+\cdots+t_mx^m$. 由 $s_n\neq 0$ 知 s_n 是一个单位. 若 $\deg(r)\geqslant\deg(f)$, 定义

$$h(x)=r(x)-t_ms_n^{-1}x^{m-n}f(x),$$

即如果 $\mathrm{LT}(f)=s_nx^n$, 其中 LT 代表末项, 则

$$h=r-\frac{\mathrm{LT}(r)}{\mathrm{LT}(f)}f,$$

注意或者 $h=0$, 或者 $\deg(h)<\deg(r)$. 如果 $h=0$, 则 $r=[\mathrm{LT}(r)/\mathrm{LT}(f)]f$ 且

$$g=qf+r=qf+\frac{\mathrm{LT}(r)}{\mathrm{LT}(f)}f=\left[q+\frac{\mathrm{LT}(r)}{\mathrm{LT}(f)}\right]f,$$

与 $f\nmid g$ 相矛盾. 如果 $h\neq 0$, 则 $\deg(h)<\deg(r)$ 且

$$g-qf=r=h+\frac{\mathrm{LT}(r)}{\mathrm{LT}(f)}f,$$

于是 $g-[q+\mathrm{LT}(r)/\mathrm{LT}(f)]f=h$, 与 r 的定义相矛盾. 所以 $\deg(r)<\deg(f)$.

下证唯一性. 设 $g=q'f+r'$, 其中 $\deg(r')<\deg(f)$, 则

$$(q-q')f=r'-r.$$

如果 $r'\neq r$, 则两端都有次数. 但 $\deg((q-q')f)=\deg(q-q')+\deg(f)\geqslant\deg(f)$, 而 $\deg(r'-r)<\deg(f)$, 于是产生矛盾. 因此 $r=r'$, 且 $(q-q')f=0$. 即 $q=q'$. □

Definition 3.5.7 An integral domain R in which every ideal is principal ideal (see Example 3.3.4) is called a **principal ideal domain**(主理想整环), often denoted by **PID**.

Example 3.5.8 $\mathbb{Z}$ is a PID, for $\mathbb{Z}$ is certainly an integral domain and if J is a subring of $\mathbb{Z}$, then $J=n\mathbb{Z}=\langle n\rangle$ for some $n\in\mathbb{Z}$.

Example 3.5.9 If F is a field, then $F[x]$ is a PID.

Definition 3.5.10 Let D be an integral domain. $0\neq q\in D$ is not a unit. If q has no proper divisors in D (i.e., the only divisors of q are units and associates of q), then q is called an **irreducible element** (不可约元) of D. If $0\neq p\in D$ is not a unit and $p\mid ab$ implies $p\mid a$ or $p\mid b$ for any $a,b\in D$, then p is called a **prime** (素元) in D.

Example 3.5.11 The primes and irreducible elements in $\mathbb{Z}$ coincide.

Example 3.5.12 An element $p \in D$ is a prime if and only if $\langle p \rangle$ is a proper prime ideal in D. Indeed, $p \neq 0$ if and only if $\langle p \rangle \neq 0$; p is not a unit if and only if $\langle p \rangle \neq D$, and $p \mid a$ if and only if $a \in \langle p \rangle$. So the statement "$p \mid ab$ implies $p \mid a$ or $p \mid b$" is equivalent to the statement "$ab \in \langle p \rangle$ implies $a \in \langle p \rangle$ or $b \in \langle p \rangle$".

Example 3.5.13 If D is PID, then an element $q \in D$ is an irreducible element if and only if $\langle q \rangle$ is a maximal ideal (see Exercise).

Theorem 3.5.14 Any prime element in D is irreducible where D is any domain.

证明 设 p 是 D 中任一素元, a 为 p 的因子, 且 $p = ab$. 进而有 $p \mid ab$. 由 p 为素元, 故 $p \mid a$ 或 $p \mid b$.

若 $p \mid a$, 设 $a = pc$, 则有 $p = pcb$, 进而 $cb = 1$, 即 b 为单位. 从而 a 与 p 相伴.

同理可得, 若 $p \mid b$, 则 b 与 p 相伴. 故 p 没有真因子, 即 p 为不可约元. □

Definition 3.5.15 An integral domain D is called a **unique factorization domain** (唯一分解整环) (abbreviate **UFD**), provided that any nonzero, nonunit element of D is a product of a finite number of primes and this factorization is unique except for order and associates. Thus if

$$\prod_{i=1}^{n} p_i = \prod_{j=1}^{m} q_j$$

and the $p_i (i = 1, 2, \cdots, n)$ and $q_j (j = 1, 2, \cdots, m)$ are primes in D, then $m = n$, and there exists $\sigma \in S_n$ such that p_i and $q_{\sigma(i)}$ are associates, $i = 1, 2, \cdots, n$.

Example 3.5.16 $\mathbb{Z}$, $F[x]$, $\mathbb{Z}[i]$ are UFD.

Example 3.5.17 Let D be a UFD, then $p \in D$ is a prime if and only if p is an irreducible element.

证明 设 p 是 D 中素元, 由定理 3.5.14 可知 p 为不可约元. 反之, 若 p 为不可约元, 且 $p \mid ab$, 设 $ab = pc$, 若 a, b 中有零元或者单位, 则显然 p 至少整除 a, b 中的一个. 因此不妨设 a 与 b 既不是零元也不是单位.

我们可知 c 不是单位, 否则, 有 $abc^{-1} = p$, 与 p 为不可约元相矛盾. 由 D 为唯一分解整环, 设 $a = q_1 q_2 \cdots q_r$, $b = u_1 u_2 \cdots u_s$, $c = p_1 p_2 \cdots p_n$, 其中 q_i, u_j, p_t 为不可约元, 则

$$q_1 q_2 \cdots q_r u_1 u_2 \cdots u_s = p p_1 p_2 \cdots p_n.$$

进而可知 p 必与某个 q_i 或者 u_j 相伴. 若 p 与 q_i 相伴, 则 $p \mid a$; 若 p 与 u_j 相伴, 则 $p \mid b$, 即 p 必然整除 a, b 中的某个. 于是 p 是素元.

综上可知 UFD 中不可约元与素元等价. □

Theorem 3.5.18 If D is a PID, then D is a UFD.

证明 若 a 为 D 中的不可约非零非单位的元素, 设 $a = a_1 b_1$. 下证 a_1 只含有有限个不可约真因子. 设 a_2 为 a_1 的真因子, a_3 为 a_2 的真因子, $\cdots$, a_{n+1} 为 a_n 的真因子, $\cdots$, 则 $\langle a \rangle \subseteq \langle a_1 \rangle \subseteq \langle a_2 \rangle \subseteq \cdots$, 令

$$I = \bigcup_{i=1}^{\infty} \langle a_i \rangle.$$

由 D 为 PID, 可设 $I = \langle d \rangle$, 则必存在 j 使得 $d \in \langle a_j \rangle$, 此时有

$$\langle a_j \rangle = \langle d \rangle = I.$$

因而 $I = \langle d \rangle = \langle a_j \rangle = \langle a_{j+1} \rangle = \cdots$, 即此时序列

$$a_1, a_2, \cdots, a_j, a_{j+1}, \cdots$$

停止于 a_{j+1}.

同理可证 b_1 也只含有有限个不可约真因子, 即 a 的真因子分解是有限个.

再设 a 有两个不可约真因子分解

$$a = p_1 p_2 \cdots p_r = q_1 q_2 \cdots q_s,$$

则易知 $p_1 \mid q_1 q_2 \cdots q_s$. 由 p_1 为素元可知 p_1 必整除 $q_1, q_2, \cdots, q_s$ 中的一个. 不妨设 $p_1 \mid q_1$. 由 q_1 亦为不可约元可知 p_1 与 q_1 相伴, 即存在单位 $u_1 \in D$, 使得 $q_1 = u_1 p_1$. 进而有

$$p_2 p_3 \cdots p_r = u_1 q_2 \cdots q_s,$$

如此继续进行下去, 若 $r \neq s$, 不妨设 $r < s$, 则

$$1 = u_1 \cdots u_r q_{r+1} \cdots q_s,$$

即说明 $u_1 \cdots u_r q_{r+1} \cdots q_{s-1}$ 与 q_s 互逆, 与 q_s 为不可约元相矛盾. 因此 $r = s$. 且经过适当调整顺序后, p_i 与 q_j 相伴. 故 D 为 UFD. □

Example 3.5.19 $\mathbb{Z}$ is a UFD since Example 3.5.8 says that $\mathbb{Z}$ is a PID.

Definition 3.5.20 The **Euclidean norm** (欧几里得模) on an integral domain D is a functor ν mapping the nonzero elements of D into the nonnegative integers such that the following conditions are satisfied:

(1) For all $a, b \in D$ with $b \neq 0$, there exist q and r in D such that $a = bq + r$, where either $r = 0$ or $\nu(r) < \nu(b)$.

(2) For all $a, b \in D$, where neither a nor b is 0, $\nu(a) \leqslant \nu(ab)$.

An integral domain is an **Euclidean domain**(欧几里得整环) if there exists an Euclidean norm on D.

Example 3.5.21 The integral domain $\mathbb{Z}$ is an Euclidean domain, for the functor ν defined by $\nu(n) = |n|$ for $n \neq 0$ in $\mathbb{Z}$ is an Euclidean norm on $\mathbb{Z}$. Condition (1) holds by the division algorithm for $\mathbb{Z}$. Condition (2) follows from $|ab| = |a||b|$ and $|a| \geqslant 1$ for $a \neq 0$ in $\mathbb{Z}$.

Example 3.5.22 If F is a field, then $F[x]$ is an Euclidean domain, for the function ν defined by $\nu(f(x)) = \deg(f)$ for $f(x) \in F[x]$, and $f(x) \neq 0$ is an Euclidean norm. Condition (1) holds by Theorem 3.5.6, and condition (2) holds since the degree of the product of two polynomials is the sum of their degrees.

Theorem 3.5.23 Every Euclidean domain D is a PID.

证明 设 I 为 D 中的理想. 若 $I = 0$, 则显然 I 是主理想. 设 $I \neq 0$. 由 D 为欧几里得整环, 知存在从 D^* 到非负整数集的一个映射 ν, 则在 I 的像中一定存在一个最小的, 设为 $\nu(a)$, 其中 $a \in I, a \neq 0$.

任取 $b \in I$, 则 $q, r \in D$, 使得

$$b = aq + r, \quad r = 0\text{或}\nu(r) < \nu(a).$$

因为 $a, b \in I$, I 为理想, 所以

$$r = b - aq \in I.$$

若 $r \neq 0$, 则 $\nu(r) < \nu(a)$, 与 $\nu(a)$ 的最小性相矛盾. 于是 $r = 0$. 从而

$$b = aq \in \langle a \rangle,$$

故 $I = \langle a \rangle$, 即 I 为主理想. □

Corollary 3.5.24 An Euclidean domain D is a UFD.

3.6 域 (Fields)

本节主要讨论域的基本性质.

Lemma 3.6.1 Let R be a ring. Then R is a division ring if and only if for any $a, b \in R$, and $a \neq 0$, the equation $ax = b$ (or $ya = b$) has solution in R.

证明 设 R 为除环, 故 R^* 为乘法群. 引理显然成立.

反之, 设方程 $ax = b$ 在 R 中有解. 首先若对于任意非零元 $a, b \in R$, 有 $ab = 0$. 不妨设

$$ac = b, \quad bd = c,$$

从而

$$abd = ac = b,$$

与 $ab = 0$ 矛盾, 即在 R 中无零因子, 从而消去律成立.

我们可知 $ax = a$ 有解, 设为 e. 对任意的 $c \in R^*$, 有 $aec = ac$. 故由消去律知 $ec = c$, 即 e 为 R^* 的左单位元. 特别地, 取 $e = c$, 得到 $e^2 = e$. 再任取 $d \in R^*$, 有 $de^2 = de$, 由消去律知 $de = d$, 即 e 亦为 R^* 的右单位元.

最后, 对任意的 $a \in R^*$, 由 $ax = e$ 有解可知 a 有右逆元. 从而 R^* 为乘群, 即 R 为除环.

同理可证当方程 $ya = b$ 在 R 中有解时, R 也为除环. □

Theorem 3.6.2 F is a commutative ring, then F is a field if and only if for any $a, b \in F$, and $a \neq 0$, the equation $ax = b$ (or $ya = b$) has solution in F.

证明 由引理 3.6.1 可直接得到. □

Theorem 3.6.3 Every finite integral domain is a field.

证明 设 $0, 1, a_1, \cdots, a_n$ 是有限整环 D 中的所有元素. 下证对 D 中任意的非零元素 a, 存在 $b \in D$ 使得 $ab = 1$. 考虑

$$a1, aa_1, \cdots, aa_n.$$

因为 D 中的元素两两不同, $aa_i = aa_j$ 意味着 $a_i = a_j$. 又 D 无零因子, 上述 $n+1$ 个元素都不是零元素. 重新排序, 不难发现 $a1, aa_1, \cdots, aa_n$ 是 $1, a_1, \cdots, a_n$ 的一种排列, 因而或者 $a = 1$, 或者存在某个 i 使得 $aa_i = 1$, 即 a 是乘法可逆的. □

Example 3.6.4 If p is prime, then $\mathbb{Z}_p$ is a field.

Let R be any ring. We might ask whether there is a positive integer n such that $na = 0$ for all $a \in R$, where $na = a + a + \cdots + a$ for n summands. For example, the integer m has this property for the ring $\mathbb{Z}_m$.

Definition 3.6.5 If for a ring R there exists a positive integer n such that $na = 0$ for all $a \in R$, then the least such positive integer is the **characteristic** (特征) of the ring R. If no such positive integer exists, R is of **characteristic 0**.

We shall be using the concept of characteristic chiefly for fields.

Example 3.6.6 The ring $\mathbb{Z}_m$ is of characteristic n, while $\mathbb{Z}, \mathbb{Q}, \mathbb{R}$ and $\mathbb{C}$ all have characteristic 0.

At first glance, determination of the characteristic of a ring seems to be a tough job, unless the ring is obviously of characteristic 0. Do we have to examine every element a of the ring in accordance with Definition 3.6.5? Our final theorem of this section shows that if the ring has unity, it suffices to examine only $a = 1$.

Theorem 3.6.7 Let R be a ring with identity. If $n1 \neq 0$ for all $n \in \mathbb{Z}^+$, then R has the characteristic 0. If $n1 = 0$ for some $n \in \mathbb{Z}^+$, then the smallest such integer n is the characteristic of R.

证明 如果对任意的 $n \in \mathbb{Z}^+$, $n1 \neq 0$, 那么不存在某个 $n \in \mathbb{Z}^+$, 对任意 $a \in R$ 都有 $na = 0$. 由定义 3.6.5,R 的特征为 0. 假设 $n \in \mathbb{Z}^+$ 使得 $n1 = 0$, 则对任意 $a \in R$, 有

$$na = a + a + \cdots + a = a(1 + 1 + \cdots + 1) = a(n1) = a0 = 0.$$

□

Theorem 3.6.8 Let R be a ring with identity and characteristic $n > 0$.

(1) If $\varphi : \mathbb{Z} \to R$ is the map given by $m \mapsto m1_R$, then φ is a homomorphism of rings with kernel $\langle n \rangle$.

(2) n is the least positive integer such that $n1_R = 0$.

(3) If R has no zero divisors (in particular if R is an integral domain), then n is prime.

证明 (1) 显然.

(2) 如果 k 是最小的正整数使得 $k1_R = 0$, 那么对任意的 $a \in R$, $ka = k(1_R a) = (k1_R)a = 0a = 0$.

(3) 如果 $n = kr, 1 < k < n, 1 < r < n$, 那么由 $0 = n1_R = (kr)1_R 1_R = (k1_R)(r1_R)$ 可得 $k1_R = 0$ 或 $r1_R = 0$, 矛盾. □

Theorem 3.6.9 Every ring R may be embedded in a ring S with identity. The ring S (which is not unique) may be chosen to be either of characteristic 0 or of the same characteristic as R.

证明 令 S 是加法交换群 $R \oplus Z$, 在 S 中定义乘法

$$(r_1, k_1)(r_2, k_2) = (r_1 r_2 + k_2 r_1 + k_1 r_2, k_1 k_2), (r_1, r_2 \in R, k_1, k_2 \in Z).$$

不难验证 S 是一个特征为 0 的有单位元 $(0,1)$ 的环, 映射 $R \to S, r \mapsto (r, 0)$ 是一个环的单同态. 如果 R 的特征为正整数 n, 类似地, 取 $S = R \oplus Z_n$, 乘法定义为

$$(r_1, \bar{k_1})(r_2, \bar{k_2}) = (r_1 r_2 + k_2 r_1 + k_1 r_2, \overline{k_1 k_2}), (r_1, r_2 \in R, \bar{k_1}, \bar{k_2} \in Z_n).$$

S 的特征为 n. □

3.7 应用 (Applications)

本节给出环的基本理论的一些应用.

Definitions 3.7.1 With each binary word $\alpha = a_0 a_1 \ldots a_{n-1}$ we associate the polynomial

$$\pi(\alpha) = \sum_{j=0}^{n-1} a_j x^j.$$

If, for example $\alpha = 1011$, then the associated polynomial is

$$\pi(\alpha) = 1 + x^2 + x^3.$$

It is easy to check that the map π is an isomorphism of the group $(\mathbf{Z}_2^n, +)$ to the subgroup $(\mathbf{Z}_2[x], +)$ consisting of all polynomials of degree less than n. If C is a subgroup of $\mathbf{Z}_2^n$, then clearly its image $\pi(C)$ is a subgroup of $\mathbf{Z}_2[x]$ isomorphic to C.

Notation The set of polynomials in $\mathbf{Z}_2[x]$ of degree less than n will be denoted by $(\mathbf{Z}_2[x]; n)$.

Definition 3.7.2 Let $p(x)$ be a fixed polynomial of degree $n - k$. The polynomial code $P_{n,k}$ generated by $p(x)$ consists of all words α of length n such that $\pi(\alpha)$ is divisible by $p(x)$. More explicitly,

$$P_{n,k} = \pi^{-1}\{c(x) \mid c(x) \in \mathbf{Z}_2[x] \text{ and } c(x) = p(x)q(x), q(x) \in (\mathbf{Z}_2[x]; k)\}.$$

Observe that there are 2^k polynomials over $\mathbf{Z}_2$ of degree at most k. Therefore $|P_{n,k}| = 2^k$.

Example 3.7.3 Let $P_{5,3}$ be the polynomial code generated by $1+x+x^2$ in the example above and let us encode $\alpha = 101$. We have $\pi(101) = 1+x^2$. Therefore $\pi(\alpha)x^2 = x^2 + x^4$. Divide this by $1+x+x^2$ to obtain

$$\pi(\alpha)x^2 = (1+x+x^2)^2 + 1.$$

Therefore the remainder is $1+0x$ and so 101 is encoded as $10\dot{:}101$. We note that the polynomial corresponding to $10\dot{:}101$ is $1+x^2+x^4$ which is divisible by $1+x+x^2$.

Example 3.7.4 Let $p(x) = 1+x+x^3, n = 5$. The generator matrix of this code is

$$\begin{pmatrix} 1 & 0 \\ 1 & 1 \\ 0 & 1 \\ 1 & 0 \\ 0 & 1 \end{pmatrix}$$

and the code words are $00000, 11010, 01101, 10111$. We note that $1+x+x^3$ does not divide $1+x^s$ for $s<5$ and therefore, as predicted $\mu(P_{5,2}) \geqslant 3$.

国际前沿研究动态

(1) 乘子环 (multiplier ring) 的研究见文献 [4].

(2) Hopf 代数的交叉积 (crossed product for Hopf algebras) 的研究见文献 [1]、[5]、[6]、[11]、[12]、[14]、[15] 和 [17]~[19].

(3) 余环 (coring) 理论的研究见文献 [3]、[7]~[10] 和 [16].

参 考 文 献

[1] Blattner R J, Cohen M, Montgomery S. Crossed products and inner actions of Hopf algebras. Tans. Amer. Math. Soc., 1986, 298(2):671–711.

[2] Brzezinski T, Wisbauer R. Corings and Comodules//Reid M, ed. London Mathematical Society Lecture Note Series. Vol.309. Cambridge: Cambridge University Press, 2003.

[3] Caenepeel S, Janssen K, Wang S H. Group coring. Appl. Categor. Struct., 2008, 16(1-2): 65-96.

[4] Dauns J. Multiplier rings and primitive ideals. Trans. Amer. Math. Soc., 1969, 145: 125-158.

[5] Delvaux L, Van Daele A, Wang S H. Bicrossproducts of multiplier Hopf algebras. J. Algebra, 2011, 343: 11-36.

[6] Elliott G A, Natsume T, Nest R. Cyclic cohomology for one-parameter smooth crossed products. Acta Math., 1988, 160(3-4): 285-305.

[7] Goodearl K R. Von Neumann Regular Rings. London: Pitman, 1979.

[8] Kac G I. Ring groups and the principle of duality, I. Trans. Moscow Math. Soc. 1963, 12: 291-339.

[9] Kac G I. Ring groups and the principle of duality, II. Trans. Moscow Math. Soc. 1965, 13: 94-126.

[10] Montgomery S. Hopf Algebras and Their Actions on Rings//AMS & CBMS, ed. CBMS Regional Conference Series in Mathematics. Vol.82. US: American Mathematical Society, 1992.

[11] Turaev V G. Homotopy field theory in dimension 3 and crossed group-categories. 2000, preprint GT/0005291.

[12] Turaev V. Crossed group-categories. Arab. J. Sci. Eng., 2008, 33(2C): 483-503.

[13] Van Daele A. The Fourier transform in quantum group theory. Proceedings of New Techniques in Hopf Algebras and Graded Ring Theory. 2007: 187-196.

[14] Van Daele A, Wang S H. New braided crossed categories and Drinfeld quantum double for weak Hopf group coalgebras. Comm. Algebra, 2008, 36(6): 2341-2386.

[15] 王栓宏. 群交叉 Yetter-Drinfel'd 范畴. 北京: 科学出版社, 2012.

[16] 王栓宏, 陈建龙. Galois 余环理论. 北京: 科学出版社, 2010.

[17] Wang S H, Jiao Z M, Zhao W Z. Hopf algebras structure on crossed products. Comm. Algebra, 1998, 26(4): 1293-1303.

[18] Zunino M. Double construction for crossed Hopf coalgebras. J. Algebra, 2004, 278: 43-75.

[19] Zunino M. Yetter-Drinfeld modules for crossed structures. J. Pure Appl. Alg., 2004, 193(1): 313-343.

习　题

1. In the following exercises, decide whether the indicated operations of addition and multiplication are defined (closed) on the set, and give a ring structure. If a ring is not formed, tell why this is the case. If a ring is formed, state whether the ring is commutative, whether it has unity, and whether it is an integral domain.

(i) $\mathbb{Z}^+$ with the usual addition and multiplication;

(ii) The set $\mathbb{R}^{\mathbb{R}}$ of all maps from the reals to the reals with operations defined as follows:

Addition: for all $f, g \in \mathbb{R}^{\mathbb{R}}$ and for all $x \in \mathbb{R}$,

$$(f+g)(x) = f(x) + g(x).$$

Multiplication: for all $f, g \in \mathbb{R}^{\mathbb{R}}$ and for all $x \in \mathbb{R}$,

$$(fg)(x) = f(x)g(x).$$

(iii) $T = \{m + n\sqrt{2} \mid m, n \in \mathbb{Z}\}$ with the usual addition and multiplication;

(iv) $T = \{m + n\sqrt{2} \mid m, n \in \mathbb{Q}\}$ with the usual addition and multiplication;

(v) The set of all pure imaginary complex numbers ri for $r \in \mathbb{R}$ with the usual addition and multiplication;

(vi) Given a set T, define on $P(T)$, the power set of T, the operations $+$ and $\cdot$ by

Addition: for all $X, Y \in P(T)$,

$$X + Y = (X \cap Y') \cup (X' \cap Y).$$

Multiplication: for all $X, Y \in P(T)$,

$$X \cdot Y = X \cap Y.$$

2. Consider the matrix ring $M_2(\mathbb{Z}_2)$, and find the **order** of the ring, that is, the number of elements in it. And list all units in the ring.

3. Let R be a ring with left unity e. Prove that if R has no zero divisors, then e is the unity of R.

4. $A \in M_n(\mathbb{R})$, then A is either a unit or a zero divisor.

5. $Z(R) = \{a \in R \mid ar = ra, \forall r \in R\}$ is called the **center** of a ring R. Prove that $Z(R)$ is a subring of R and the center of a division ring is a field.

6. Give an example of a homomorphism $f : R \to S$ where R and S are rings with unity, and where $f(1_R) \neq 0_S$ and $f(1_R) \neq 1_S$.

7. Consider the map det of $M_n(\mathbb{R})$ into $\mathbb{R}$ where $\det(A)$ is the determinant of the matrix $A \in M_n(\mathbb{R})$. Is det a ring homomorphism? Why or why not?

8. $f : R \to S$ is a ring homomorphism. If r is a unit in R, then $f(r)$ is a unit in S.

9. Describe all ring homomorphisms of $\mathbb{Z}$ into $\mathbb{Z}$.

10. Describe all ring isomorphisms of $\mathbb{Z}$ into $\mathbb{Z}$.

11. Describe all ring homomorphisms of $\mathbb{Z}_{15}$ into $\mathbb{Z}_3$.

12. Prove Theorem 3.4.6.

13. Show that if R is a ring with unity, I is a proper ideal of R, then I contains no units.

14. An element a of a ring R is **nilpotent** if $a^n = 0$ for some $n \in \mathbb{Z}^+$. Show that the collection of all nilpotent elements in a commutative ring R is an ideal, the **nilradical** of R.

15. Find all prime and maximal ideals of $\mathbb{Z}_{12}$.

16. If A and B are ideals of a ring R, the **sum** $A + B$ of A and B is defined by

$$A + B = \{a + b \mid a \in A, b \in B\},$$

and the **product** AB of A and B is defined by

$$AB = \{ab \mid a \in A, b \in B\}.$$

Show that $A + B$ and AB are ideals of R.

17. Show that the set S of all matrices of the form

$$\begin{pmatrix} a & b \\ 0 & 0 \end{pmatrix}$$

for $a, b \in \mathbb{R}$ is a right ideal but not a left ideal of $M_2(\mathbb{R})$.

18. Show that I is a maximal ideal in a ring if and only if R/I is a **simple ring**, that is, it is nontrivial and has no proper nontrivial ideals.

19. Show that $\mathbb{R}[x]/\langle x\rangle \cong \mathbb{R}$ where $\langle x\rangle$ means the principal ideal generalized by x. And furthermore, $\langle x\rangle$ is a maximal ideal of $\mathbb{R}[x]$.

20. Show that the **Gauss domain** $\mathbb{Z}[i] = \{a + bi \mid a, b \in \mathbb{Z}[i]\}$ is isomorphic to $\mathbb{Z}[x]/\langle x^2 + 1\rangle$.

21. Let F be a field and $f(x), g(x) \in F[x]$. Show that $f(x)$ divides $g(x)$ if and only if $g(x) \in \langle f(x)\rangle$.

22. Prove that if F is a field, every proper nontrivial prime ideal of $F[x]$ is maximal.

23. R is an integral domain. Prove that $f(x) \in R[x]$ is a unit if and only if $f(x) \in R$ and is a unit in R.

24. If a associates of b in a domain D, and b is an irreducible element, then a is also an irreducible element.

25. Prove Example 3.5.13.

26. Assume that D is an integral domain with unity, if D satisfies

(i) every nonzero, nonunit element is a product of irreducible elements;

(ii) every irreducible element in D is prime.

Then D is UFD.

27. Let D be a PID. If the sequence

$$a_1, a_2, \cdots, a_n, \cdots$$

for any $a_i \in D$ satisfies a_i is a proper divisor of a_{i-1}, then the sequence is finite.

(Hint: Consider the ideal $I = \langle a_1\rangle \cup \langle a_2\rangle \cup \cdots$ is a principal ideal)

28. Let $F = \{0, 1, a, b\}$ be a field. Then

(i) its characteristic is 2;

(ii) a, b satisfy $x^2 = x + 1$.

29. Show that 1 and $p - 1$ are the only elements of the field $\mathbb{Z}_p$ that are their own multiplicative inverse.

(Hint: Consider the equation $x^2 - 1 = 0$)

30. Prove that if a ring R has no proper nontrivial ideals, then its characteristic is either 0 or prime.

第4章　模 (Modules)

本章主要介绍模 (module) 的概念及其他的结构性质, 涉及子模 (submodule)、模同态和商模 (quotient module). 它的深层次研究会涉及 Hopf 模 (Hopf modules) 与余模 (comodule) 理论的建立. 本章的研究与学习方法与线性代数的理论建立方法相同.

The concept of a module is an immediate generalization of that of a vector space. One obtains the generalization by simply replacing the underlying field by any ring. Why dose one make this generalization? In the first place, one learns from experience that the internal logical structure of mathematics strongly urges the pursuit of such natural generalizations. These often results in an improved insight into the theory which led to them in the first place. A good illustration of this is afforded by the study of a linear transformation in a finite dimensional vector space over a field—a central problem of linear algebras.

Modules first became an important tool in algebra in the late 1920's, largely due to the insight of Emmy Noether, who was the first to realize the potential of the module concept. In particular, she observed that this concept could be used to bridge the gap between two important developments in algebra that had been going on side by side and independently: the theory of homomorphisms of finite groups by matrices due to Frobenius, Burnside, and Schur, as well as the structure theory of algebras due to Molien, Cartan, and Wedderburn.

4.1　模的定义与例子
(Definitions and Examples of Modules)

Modules over a ring are a generation of Abelian groups (which are modules over $\mathbb{Z}$). Consequently, this section is primarily concerned with carrying over to modules various concepts and results of group theory.

Definition 4.1.1 Let R be a ring. A left R**-module** (R-模) is an additive Abelian group M together with a map (also called **scalar multiplication**) (标量乘法) $R \times M \to M$ (the image of (r, a) denoted by ra) such that for all $r, s \in R$ and $m, n \in M$:

(1) $r(m + n) = rm + rn$;

(2) $(r + s)m = rm + sm$;

(3) $r(sm) = (rs)m$.

If R has an identity element 1_R and

(4) $1_R m = m$ for all $m \in M$,

then M is said to be a **unitary** R**-module**. If R is a division ring, then a unitary R-module is called a **vector space** (向量空间).

Similarly, one can define right R-modules. From now on, unless specified otherwise, "R-module" means "left R-module" and it is understood that all theorems about left R-modules also hold for right R-modules.

A given group M may have many different R-module structures. If R is commutative, it is easy to verify that every left R-module M can be given the structure of a right R-module M by defining $rm = mr$ for $r \in R$ and $m \in M$ (commutativity is needed for (3)). More generally, every ideal I in R is an R-module.

If M is a module with additive identity element 0_M over a ring R with additive identity element 0_R, then it is easy to show that for $r \in R$ and $m \in M$:

$$r0_M = 0_M \text{ and } 0_R m = 0_M.$$

In the sequel $0_M, 0_R, 0 \in Z$ and the trivial module $\{0\}$ will be denoted by 0.

It is also easy to verify that for all $r \in R, n \in Z$ and $m \in M$:

$$(-r)m = -(rm) = r(-m) \text{ and } n(rm) = r(nm),$$

where nm has its usual meaning for groups.

Here are some important examples of modules.

Example 4.1.2 Let $R = F$ be a field and $M = V$ be a vector space over F. Then V is an additive Abelian group. Now the map: $F \times V \to V$ via $(a, x) \mapsto ax$ satisfies (1)~(4). Thus V is a unitary R-module.

Example 4.1.3 Every additive Abelian group G is a unitary $\mathbb{Z}$-module, with $ng(n \in \mathbb{Z}, g \in G)$ given by usual meaning for groups.

Example 4.1.4 Let $M = V$ be a vector space over a field F, and $R = F[x]$ be a polynomial ring and T be a linear transformation on V. The vector space V can be made into a $F[x]$-module by defining a scalar multiplication $F[x] \times V \to V$ as follows:

$$(f(x), v) \mapsto f(T)(v).$$

Then V is an $F[x]$-module. If $f(x) = \sum_0^n a_i x^i$ lies in $F[x]$, then $f(T) = \sum_0^n a_i T^i$ and $f(x)v = \sum_0^n a_i T^i(v)$, where T^0 is the identity map 1_V, $T^1 = T$, and T^i denotes the composite of T with itself i times if $i \geqslant 2$.

Example 4.1.5 Let G be an additive Abelian group. Then G is an $End(G)$-module, where scalar multiplication $End(G) \times G \to G$ is defined by $(f, a) \mapsto f(a)$. Let us check the associativity axiom (3). We use extra -fussy notation here: write $f \circ g$ to denote the composite (which is the product of f and g in $End(G)$), and write $f * a$ to denote the action of f on a (so that $f * a = f(a)$). Now $(fg) * a = (f \circ g) * a = f(g(a))$, while $f * (g * a) = f * (g(a)) = f(g(a))$. Thus $(fg) * a = f * (g * a)$.

4.2 子模 (Submodules)

本节主要介绍子模 (submodule) 的概念与循环模的结构性质.

Definition 4.2.1 Let R be a ring, M a R-module and N a nonempty subset of M. N is a **submodule**(子模) of M provided that N is an additive Abelian subgroup of M and $rn \in N$ for all $r \in R$ and $n \in N$, denoted by $N \leqslant M$. A submodule of a vector space over a division ring is called a **subspace**.

Note that a submodule is itself a module. Also a submodule of a unitary module over a unitary ring is necessarily unitary.

Example 4.2.2 Both $\{0\}$ and M are submodules of a R-module, and are called **trivial submodules**(平凡子模). A **proper submodule**(真子模) of M is a submodule $N \leqslant M$ with $N \neq M$. In this case, we may write $N \leqslant M$.

Example 4.2.3 If a ring R is viewed as a left R-module, then a submodule of R is a left ideal; I is a proper submodule when it is a proper ideal.

Example 4.2.4 A submodule of a $\mathbb{Z}$-module (i.e., of an Abelian group) is a subgroup, and a submodule of a vector space is a subspace.

Example 4.2.5 If $(S_i)_{i\in I}$ is a family of a R-module M, then $\cap_{i\in I}S_i$ is a submodule of M.

Definition 4.2.6 If X is a subset of a R-module M, then the intersection of all submodules of M containing X is called the **submodule generated** by X or **spanned by X**.

Note that if X is finite, and X generates the R-module N, N is said to be **finitely generated**(有限生成的). If $X=\phi$, then X clearly generates the trivial R-module. If X consists of a single element, that is, $X=\{a\}$, then the submodule generated by X is called the **cyclic (sub)module** generated by a. Finally, if $(S_i)_{i\in I}$ is a family of a R-module M, then the submodule generated by $X=\cup_{i\in I}S_i$ is called the **sum** of the modules S_i. If the index set I is finite, the sum of $S_1, S_2, \cdots, S_n$ is denoted by $S_1+S_2+\cdots+S_n$.

Theorem 4.2.7 Let R be a ring, M a R-module and X a subset of M, and $(S_i)_{i\in I}$ is a family of a submodule of M and $m\in M$. Let $Rm=\{rm|r\in R\}$, then

(1) Rm is a submodule of M;

(2) the cyclic submodule C generated by m is $\{rm+nm|r\in R, n\in Z\}$. If R has an identity and C is unitary, then $C=Rm$;

(3) the submodule D generated by X is

$$\left\{\sum_{i=1}^{s} r_i m_i+\sum_{j=1}^{t} n_j v_j | s,t\in N^*; m_i, v_j\in M; r_i\in R; n_j\in Z\right\}.$$

If R has an identity and M is unitary, then

$$D=RX=\left\{\sum_{i=1}^{s} r_i m_i | s\in N^*; m_i\in M; r_i\in R\right\};$$

(4) the sum of the family $(S_i)_{i\in I}$ consists of all finite sums $s_{i_1}+s_{i_2}+\cdots+s_{i_n}$ with $s_{i_k}\in S_{i_k}$.

证明 留给读者作为练习. 提示: 如果 R 有单位 1_R, M 是一个具有单位元的模, 那么对任意的 $m\in M, n\in Z$ 有 $n1_R\in R, nm=(n1_R)m$. □

4.3 模同态 (Module Homomorphism)

本节介绍模同态的概念和相关性质.

Definition 4.3.1 Let M and N be R-modules. A map $f : M \to N$ is a R-**module homomorphism** provided that for all $m, m' \in M$ and $r \in R$:

$$f(m+m') = f(m) + f(m') \text{ and } f(rm) = rf(m).$$

If R is a division ring, then a R-module homomorphism is called a **linear transformation** (线性变换).

Note that a R-module homomorphism $f : M \to N$ is necessarily a homomorphism of additive Abelian groups. Consequently, the same terminology is used: f is a R-**module monomorphism** [resp. **epimorphism, isomorphism**] if it is injective [resp. surjective, isomorphism] as a map of sets.

Example 4.3.2 For any R-modules M and N, the zero map $0 : M \to N$ given by $m \mapsto 0 (m \in M)$ is a R-module homomorphism.

Example 4.3.3 Every homomorphism of Abelian groups is a $\mathbb{Z}$-module homomorphism.

Example 4.3.4 If R is a ring, the map $R[x] \to R[x]$ given by $f \mapsto xf$ (for example, $(x^2+1) \mapsto x(x^2+1)$) is a R-module homomorphism, but not a ring homomorphism.

Definition 4.3.5 If $f : M \to N$ is a R-module homomorphism, then its **kernel** is

$$\ker(f) = \{m \in M | f(m) = 0\}$$

and its **image** is

$$\mathrm{im}(f) = \{n \in N | \text{ there exists } m \in M \text{ with } n = f(m)\}.$$

Proposition 4.3.6 Let $f : M \to N$ be a R-module homomorphism, then $\ker(f)$ is a submodule of M, and $\mathrm{im}(f)$ is a submodule of N.

证明 我们仅证明 $\ker(f)$ 是 M 的一个子模, $\mathrm{im}(f)$ 是 N 的子模的证明留作练习. 由命题 2.3.4 知 $\ker(f)$ 是 M 的一个交换子群. 对任意的 $m \in \ker(f), r \in R$,

因为

$$f(rm) = rf(m) = r0 = 0,$$

所以 $rm \in \ker(f)$, $\ker(f)$ 是 M 的一个子模.

□

Note that a R-module homomorphism $f : M \to N$ is a R-module monomorphism if and only if $\ker(f) = 0$; f is a R-module isomorphism if and only if there is a R-module homomorphism $g : N \to M$ such that $gf = 1_M$ and $fg = 1_N$.

4.4 商模 (Quotient Modules)

Let M be a R-module and N a submodule of M. We consider the quotient group

$$\overline{M} = M/N = \{\bar{x} = x + N | x \in M\}$$

with the addition

$$(x_1 + N) + (x_2 + N) = x_1 + x_2 + N,$$

the zero element N and $-(x+N) = -x + N$. If $\bar{x_1} = \bar{x_2}$, that is, $x_1 - x_2 \in N$, then

$$rx_1 - rx_2 = r(x_1 - x_2) \in N,$$

so $\overline{rx_1} = \overline{rx_2}$. It follows that if we put

$$r\bar{x} = r(x + N) = rx + N = \overline{rx},$$

then this coset is independent of the choice of the element x in the coset. Hence $(r, \bar{x}) \mapsto r\bar{x}$ is a map of $R \times \overline{M} \to \overline{M}$. We also have

$$r(\bar{x_1} + \bar{x_2}) = r(\overline{x_1 + x_2}) = \overline{rx_1 + rx_2} = r\bar{x_1} + r\bar{x_2},$$

$$(r + r')\bar{x} = r\bar{x} + r'\bar{x}, (rr')\bar{x} = r(r'\bar{x}), 1\bar{x} = \overline{1x} = \bar{x},$$

for all $x, x_1, x_2 \in M$ and $r, r' \in R$. Thus $\overline{M} = M/N$ with the action defined above is a R-module.

Definition 4.4.1 Let M be a R-module and N a submodule of M, the R-module $\overline{M} = M/N$ defined above is called the **quotient module** of M with respect to the submodule N.

Theorem 4.4.2 Let M be a R-module and N a submodule of M. Then the canonical map $\pi : M \to M/N$ given by $m \mapsto m + N$ is a R-module epimorphism with kernel N.

证明 对任意的 $x, y \in M, r \in R$, 因为

$$\pi(x+y) = (x+y) + N = \overline{(x+y)} = \bar{x} + \bar{y} = \pi(x) + \pi(y),$$
$$\pi(rx) = rx + N = \overline{rx} = r\bar{x} = r\pi(x),$$

所以 π 是一个 R–模同态. 又 π 是一个满射, 所以 π 是一个满的 R–模同态. $\ker \pi = N$ 是显然的.

□

4.5 模的同态基本定理

In view of the preceding results it is not surprising that the various isomorphism theorems for groups are valid, and mutatis mutandis, the same is the for modules.

Theorem 4.5.1 (The First Isomorphism Theorem) If $f : M \to N$ is a R-module homomorphism, then there is a R-module isomorphism $\varphi : M/\ker(f) \to \mathrm{im}\, f$ given by $m + \ker(f) \mapsto f(m)$.

证明 如果我们仅把 M, N 看成两个交换群, 那么群的第一同构定理表明 $\varphi : M/\ker(f) \to \mathrm{im}(f)$ 是一个交换群同构. 因为 f 是一个 R–模同态, 所以 $f(rm) = rf(m) = r\varphi(m + \ker(f))$, 从而 $\varphi(r(m + \ker(f))) = \varphi(rm + \ker(f)) = f(rm)$, φ 是一个 R–模同构.

□

Theorem 4.5.2 (The Second Isomorphism Theorem) If S, T are submodules of a R-module M, then there exists a R-module isomorphism $S/S \cap T \to (S+T)/T$.

证明 令 $\pi : M \to M/T$ 是典范同态, 则 $\ker(\pi) = T$. 定义 $h = \pi|_S$, 即 $h : S \to M/T$. 因为 $\ker(h) = S \cap T, \mathrm{im}(h) = (S+T)/T$, 由第一模同构基本定理可得 $S/S \cap T$ 与 $(S+T)/T$ 同构.

□

Theorem 4.5.3 (The Third Isomorphism Theorem) If $T \subseteq S \subseteq M$ is a tower of submodules of a R-module M, then S/T is a submodule of M/T and there exists a R-module isomorphism $(M/T)/(S/T) \to M/S$.

证明 定义映射 $g: M/T \to M/S, m+T \mapsto m+S$. 如果 $m+T=m'+T$, 那么 $m-m' \in T \subseteq S$, 从而 $m+S=m'+S$, 映射 g 是有意义的. 因为 $\ker(g)=S\cap T, \mathrm{im}(g)=M/S$, 由第一模同构基本定理可得 $(M/T)/(S/T)$ 与 M/S 同构.

□

Note that the cyclic modules are analogous to the cyclic groups. Recall that M is a R-module if there exists $m \in M$ such that $M = Rm$ and noted by $M = \langle m\rangle$. For example, cyclic group $G=\langle g\rangle$ is a cyclic $\mathbb{Z}$-module, $G=\mathbb{Z}g$. When R is a ring with identity, R is also a cyclic R-module $R=R1$.

Proposition 4.5.4 A R-module M is cyclic if and only if $M\cong R/I$ for some ideal I.

证明 如果 M 是循环的, 则存在 $m\in M$ 使得 $M=\langle m\rangle$. 定义映射 $f:R\to M, r\mapsto rm$. 因为 M 是循环的, 所以 f 是满射. 令 $I=\ker(f)$ 是 R 的一个理想, 由第一模同构基本定理,$R/I\cong M$. 反之,R/I 是由 $1+I$ 生成的循环模, 任意与循环模同构的模也是循环模.

□

Let M be a R-module, then for any $m\in M$, Rm is a submodule of M. Consider $f:R\to Rm, r\mapsto rm$, then by Theorem 4.5.1, $Rm\cong R/ker(f)$, where $\ker(f)=\{r\in R|rm=0\}$ which is called the **annihilator**(零化子) of m and denoted by $ann(m)$. So we have $Rm\cong R/ann(m)$. If $ann(m)=0$, then $Rm\cong R$.

Example 4.5.5 Let G be an Abelian group, $g\in G$. Then G is a $\mathbb{Z}$-module, $ann(m)\lhd\mathbb{Z}$. Set $ann(m)=(n)$, then $\mathbb{Z}/(n)\cong\langle g\rangle$. If $n=0$, then $ann(m)\cong\mathbb{Z}$ is an infinite cyclic group; if $n>0$, then $\langle g\rangle$ is a cyclic group of order n.

Example 4.5.6 Let R be a ring with identity, then R is a left R-module. Suppose that $\eta\in End_R(R)$, set $\eta(1)=b_\eta$, for all $r'\in R$,

$$\eta(r')=\eta(r'1)=r'\eta(1)=r'b_\eta.$$

Thus $\eta=(b_\eta)_r$ is the right multiplication transformation by b_η. Consequently, every right multiplication transformation by b_r is a R-module homomorphism. In fact,

$$b_r(ar')=(ar')b=a(r'b)=ab_r(r').$$

Hence $End_R(R)=$ the ring of transformation by right multiplication of R.

4.6 应用 (Applications)

In this section we shall apply module to the linear algebras.

Let V be a finite dimensional vector space over a field F and let $\lambda \in L(V)$. We know that V can be considered as an $F[x]$-module, with scalar multiplication defined by $f(x) \cdot v = f(\lambda)(v)$ for any $v \in V$ and $f(x) \in F[x]$. Then V is an $F[x]$-module, denoted by V_λ.

Let $\{v_1, v_2, \cdots, v_n\}$ be a basis for V, then $\{v_1, v_2, \cdots, v_n\}$ is a set of generators of V_λ, that is, V_λ is a finitely generated $F[x]$-module. Next it is easy to show that V_λ is a torsion module. For any $v \in V_\lambda$, then $ann(v) = (m(\lambda))$, where $m(\lambda)$ is the minimal polynomial of element v. Then V_λ can be decomposed into a direct sum of cyclic submodules:

$$V_\lambda = F[\lambda]z_1 \oplus F[\lambda]z_2 \oplus \cdots \oplus F[\lambda]z_t$$

such that

$$ann(z_i) = (d_i(\lambda)),$$

and

$$(d_1(\lambda)) \supseteq (d_2(\lambda)) \supseteq \cdots \supseteq (d_s(\lambda)), d_i(\lambda) \neq 0.$$

Theorem 4.6.1 (1) Every cyclic submodule $V_i = F[\lambda]z_i$ of V_λ is invariant subspace of the vector space V under λ;

(2) $\dim V_i = deg(d_i(\lambda)) = n_i$ and $\{z_i, \lambda z_i, \cdots, \lambda^{n_i-1}z_i\}$ is a basis for V_i;

(3) Let $d_i(\lambda) = \lambda^{n_i} + b_{in_i-1}\lambda^{n_i-1} + \cdots + b_{i0}$, the matrix relative to the basis $\{z_i, \lambda z_i, \cdots, \lambda^{n_i-1}z_i\}$ is

$$B_i = \begin{pmatrix} 0 & 0 & \cdots & 0 & -b_{i0} \\ 1 & 0 & \cdots & 0 & -b_{i1} \\ 0 & 1 & \cdots & 0 & -b_{i2} \\ \vdots & \vdots & & \vdots & \vdots \\ 0 & 0 & \cdots & 0 & -b_{in_i-1} \end{pmatrix}$$

where B_i is called the companion matrix of the polynomial $d_i(\lambda)$.

If the invariant factors $d_i(\lambda)$ of V_λ can be factored as products of linear factors $\lambda - r$ in $F[\lambda]$, then there is a second canonical form for a linear transformation λ, the so-called Jordan form. This is always the case if F is the field of complex number $\mathcal{C}$. Suppose that $d_s(\lambda)$ can be factored into

$$d_s(\lambda) = \prod_{i=1}^{r}(\lambda - \lambda_j)^{e_{s_j}}, e_{s_j} \geqslant 1, \lambda_i = \lambda_j, i \neq j.$$

Then $d_1(\lambda), \cdots, d_{s-1}(\lambda)$ have the form of

$$d_i(\lambda) = \prod_{j=1}^{r}(\lambda - \lambda_j)^{e_{i_j}}, e_{i_j} \geqslant 0.$$

Since $d_i(\lambda) \mid d_{i+1}(\lambda)$, we have

$$0 \leqslant e_{1_j} \leqslant e_{2_j} \leqslant \cdots \leqslant e_{s_j}, j = 1, 2, \cdots, r.$$

Then $F[\lambda]z_i$ can be decomposed into the direct sum of cyclic sub modules:

$$F[\lambda]z_i = \bigoplus_{j=1}^{r} = F[\lambda]z_{i_j}$$

where

$$ann(z_{i_j}) = (\lambda - \lambda_j)^{e_{i_j}}, e_{i_j} \geqslant 1.$$

Hence

$$V_\lambda = \bigoplus_{i}\bigoplus_{j} F[\lambda]z_{i_j}.$$

Now let $V_1 = F[\lambda]$ be a cyclic direct summand with

$$ann(z_{i_j}) = ((\lambda - \lambda_1)^e), e \geqslant 1.$$

Then F-space V_1 has the basis

$$\{z, (\lambda - \lambda_1)z, \cdots, (\lambda - \lambda_1)^{e-1}z\},$$

and we have

$$\begin{aligned}
\lambda z &= \lambda_1 z + (\lambda - \lambda_1)z,\\
(\lambda - \lambda_1)z &= \lambda_1(\lambda - \lambda_1)z + (\lambda - \lambda_1)^2 z,\\
&\vdots\\
(\lambda - \lambda_1)^{e-2}z &= \lambda_1(\lambda - \lambda_1)^{e-2}z + (\lambda - \lambda_1)^{e-1}z,\\
(\lambda - \lambda_1)^{e-1}z &= \lambda_1(\lambda - \lambda_1)^{e-1}z.
\end{aligned}$$

Therefore the matrix of the restriction of λ to $V_1 = F[\lambda]z$ relative to the basis

$$\{z, (\lambda-\lambda_1)z, \cdots, (\lambda-\lambda_1)^{e-1}z\}$$

is

$$\begin{pmatrix} \lambda_1 & 0 & \cdots & 0 & 0 \\ 1 & \lambda_1 & \cdots & 0 & 0 \\ 0 & 1 & \cdots & 0 & 0 \\ \vdots & \vdots & & \vdots & \vdots \\ 0 & 0 & \cdots & 0 & \lambda_1 \end{pmatrix}.$$

This matrix is referred to as a Jordan block associated to the scalar λ_i.

We take the basis

$$\{z_{i_j}, (\lambda-\lambda_j)z_{i_j}, \cdots, (\lambda-\lambda_j)^{e_{i_j}-1}z_{i_j}\}$$

for every cyclic submodule $F[\lambda]z_{i_j}$ and obtain a basis for V over F by stringing together bases of this types. The matrix of λ relative to this basis the Jordan canonical form

$$C = \begin{pmatrix} C_1 & 0 & \cdots & 0 & 0 \\ 1 & C_2 & \cdots & 0 & 0 \\ 0 & 1 & \cdots & 0 & 0 \\ \vdots & \vdots & & \vdots & \vdots \\ 0 & 0 & \cdots & 0 & C_s \end{pmatrix}$$

where

$$C_i = \begin{pmatrix} \lambda_i & 0 & \cdots & 0 & 0 \\ 1 & \lambda_i & \cdots & 0 & 0 \\ 0 & 1 & \cdots & 0 & 0 \\ \vdots & \vdots & & \vdots & \vdots \\ 0 & 0 & \cdots & 0 & \lambda_i \end{pmatrix}.$$

国际前沿研究动态

(1) 相关研究涉及 Hopf 模 (Hopf modules) 结构理论, 研究见文献 [1]、[3]~[12] 和 [14].

(2) 相关研究涉及余模 (comodules) 理论, 研究见文献 [2] 和 [13].

参考文献

[1] Böhm G. Doi-Hopf modules over weak Hopf algebras. Comm. Algebra. 2000, 28: 4687-4698.

[2] Brzezinski T, Wisbauer R. Corings and Comodules//London Math. Soc. Lecture Note Series. Vol.309. Cambridge: Cambridge University Press, 2003.

[3] Caenepeel S, Militaru G, Zhu S L. Frobenius and Separable Functors for Generalized Module Categories and Nonlinear Equations//Morel, et al. Lecture Notes in Mathematics. Vol.1787. Berlin: Springer Verlag, 2002.

[4] Caenepeel S, Wang D G, Yin Y M. Yetter-Drind modules over weak Hopf algebras and the center construction. Ann. Univ. Ferrara-Sez.VII-Sc. Mat., 2005, 51: 69-98.

[5] Delvaux L. On the modules of a Drinfel'd double multiplier Hopf (*-) algebras. Comm. Algebra, 2005, 33(8): 2771-2787.

[6] Delvaux L. Yetter-Drinfel'd modules for group-cograded multiplier Hopf algebras. Comm. Algebra, 2008, 36: 2872-2882.

[7] Kan H, Wang S H. A categorical interpretation of Yetter-Drinfel'd modules. Chinese Science Bulletin, 1999, 44(9): 771-778.

[8] Menini C, Militaru G. Integrals, quantum Galois extensions, and the affineness criterion for quantum Yetter-Drinfel'd modules. J. Algebra, 2002, 247(2): 467-508.

[9] Panaite F, Van Oystaeyen F. Quasi-elementary H-Azumaya algebras arising from generalzed (anti) Yetter-Drinfel'd modules. Applied Categorical Structures, 2011, 19(5): 803-820.

[10] Panaite F, Staic M D. Generalized (anti) Yetter-Drinfel'd modules as components of a braided T-category. Isr. J. Math., 2007, 158: 349-365.

[11] Schauenburg P, Hopf modules and Yetter-Drinfel'd modules. J. Algebra, 1994, 169: 874-890.

[12] Staic Mihai D. A note on Anti-Yetter-Drinfel'd modules, Contemp. Math., 2007, 441: 149-153.

[13] 王栓宏, 陈建龙. Galois 余环理论. 北京: 科学出版社, 2010.

[14] Zunino M. Yetter-Drinfeld modules for Turaev crossed structures. J. Pure and Appl. Algebra, 2004, 193: 313-343.

习　题

1. Let R be a ring, M a R-module and η a homomorphism of a ring S into R. Show that M becomes a S-module if we define $sm = \eta(s)m$ for $s \in S$ and $m \in M$.

2. Let R be a ring, A an Abelian group and $n > 0$ an integer such that $na = 0$ for all $a \in A$. Then A is a unitary Z_n-module, with the action of Z_n on A by $\bar{k}a = ka$, where $k \in Z$ and $k \mapsto \bar{k} \in Z_n$ under the canonical projection $Z \to Z_n$.

3. Show that a finitely generated R-module needs not be finitely generated as an Abelian group.

4. Let R be a ring and $f : M \to N$ a R-module homomorphism.

(i) f is a monomorphism if and only if for every pair of R-module homomorphisms $g, h : P \to M$ such that $fg = fh$, we have $g = h$. [Hint: To prove ($\Leftarrow$), let $P = \ker f$, with g the inclusion map and h the zero map]

(ii) f is an epimorphism if and only if for every pair of R-module homomorphisms $k, t : N \to Q$ such that $kf = tf$, we have $k = t$. [Hint: To prove ($\Leftarrow$), let k be the canonical epimorphism $N \to N/im\ f$ and t the zero map]

5. Let R be a ring, M, N two R-modules and $Hom_R(M, N)$ the set of all R-module homomorphisms $M \to N$.

(i) $Hom_R(M, N)$ is an Abelian group with $f+g$ given on $m \in M$ by $(f+g)(m) = f(m) + g(m) \in N$. The identity element is the zero map.

(ii) $Hom_R(M, M)$ is a ring with identity, where multiplication is composition of functions. $Hom_R(M, M)$ is called **endomorphism ring** of M.

(iii) M is a $Hom_R(M, M)$-module with fm defined by $f(m)$ for all $m \in M$ and $f \in Hom_R(M, M)$.

6. Determine $Hom(Z, Z/(N))$ and $Hom(Z/(N), Z)$, $n > 0$ (as Z-modules).

7. Let R be a ring and M a R-module, show that $Hom(R, M) \cong (M, +, 0)$.

8. Let R be a ring with identity, then a nonzero unitary R-module M is **simple** if its only submodules are 0 and M. Then every simple R-module is cyclic.

9. Show **Schur's Lemma**: Let $\varphi : S \to S'$ be a homomorphism of simple R-modules, then φ is either the zero or an isomorphism. Hence show that $End_R M$ is a division ring if M is a simple R-module.

10. Does the converse of the last statement in Exercise 9 hold: if $End_R M$ is a division ring, is M necessarily simple?

11. Let R be a ring, show that a R-module M is simple if and only if $M \cong R/I$, where I is a maximal left ideal of R.